Germany's Triplane Craze

A Centennial Perspective on Great War Airplanes

Jack Herris

Great War Aviation Centennial Series #7

This book is dedicated to the Aeronaut Team that makes this series possible; Aaron Weaver, Bob Pearson, Greg VanWyngarden, and Colin Owers. Thank you all!

ACKNOWLEDGEMENTS

My sincere thanks to Colin Owers for scale drawings and invaluable help with photographs, Aaron Weaver for photographs, and Greg VanWyngarden for photographs and his helpful suggestions and material. Thanks also to Bob Pearson for his color profiles, Aaron Weaver for digital photo editing, and Kenny Lisle, the Museum of Flight in Seattle, and the Deutsches Technikmuseum in Berlin for photographs. Any errors are my responsibility.

Cover painting by Russell Smith. Please see his website: **www.russellsmithart.com**

Color aircraft profiles © Bob Pearson. Purchase his CD of WWI aircraft profiles for $50 US/Canadian, 40 €, or £30, airmail postage included, via Paypal to Bob at: **bpearson@kaien.net**

Scale Drawings by Colin Owers. Colin Owers acknowledges the research by W Kerka for his Schütte-Lanz Dr.I drawing.

For our aviation books in print and electronic format, please see our website at: **www.aeronautbooks.com**. You may contacted me at **jherris@verizon.net**.

For our aviation books in print and electronic format, please see our website at: **www.aeronautbooks.com**.

I am looking for photographs of the less well-known German aircraft of WWI. For questions or to help with photographs you may contact me at **jherris@verizon.net**.

ISBN: **978-1-935881-15-5**

Design and layout: Jack Herris

Cover design: Aaron Weaver

Cover painting: Russell Smith

Digital photo editing: Jack Herris & Aaron Weaver

Aeronaut Books

www.aeronautbooks.com

Table of Contents

Above: Fokker F.I 101/17, the first pre-production prototype of the Dr.I, in flight at the factory at Schwerin. This aircraft was later used for the structural tests on the airframe that started 7 August 1917.

Below: Although the triplane craze was about fighters, not bombers, your humble illustrator and author could not resist sharing this classic example of triplane bomber design, the Austro-Hungarian Lloyd 40.08 prototype.

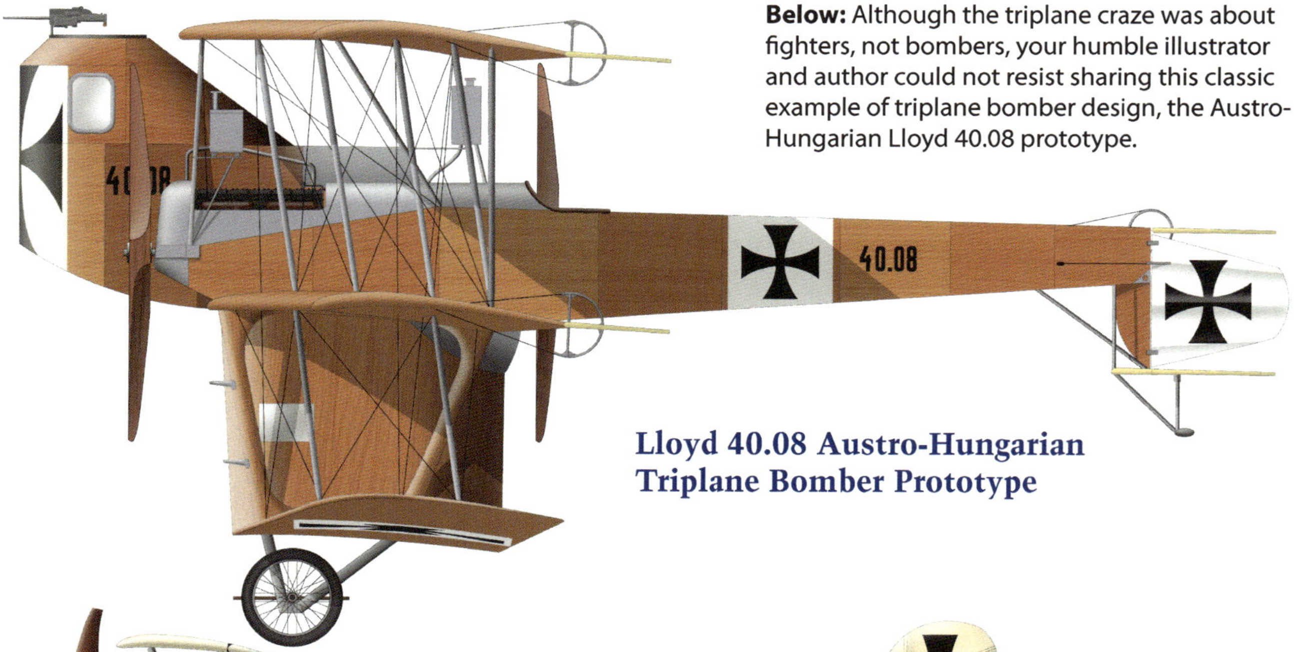

Lloyd 40.08 Austro-Hungarian Triplane Bomber Prototype

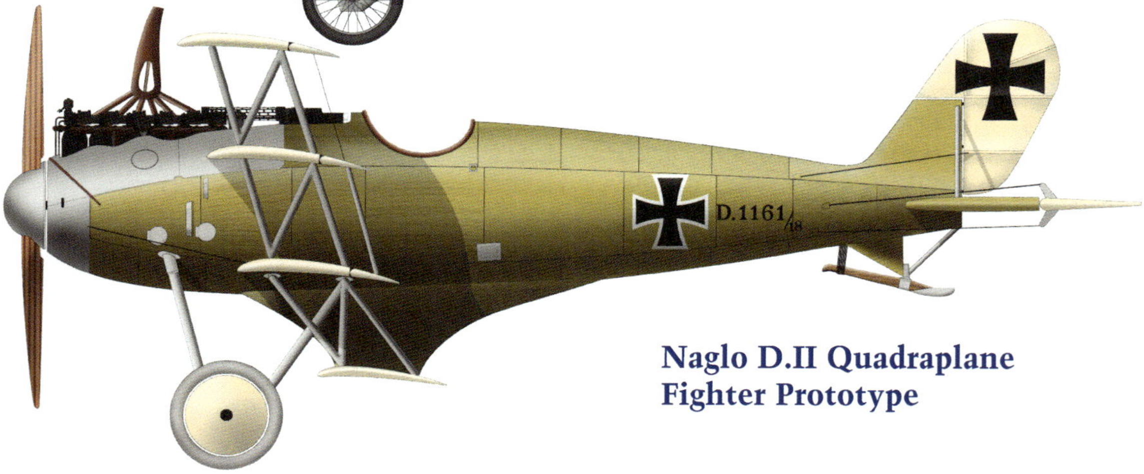

Naglo D.II Quadraplane Fighter Prototype

Prelude to Futility

Before *Idflieg's* basically futile exploration of the triplane configuration fostered by the excellent performance of the enemy's Sopwith Triplane, *Idflieg* had gone on a similar wild goose chase based on the superiority of the Nieuport fighter over the Fokker and Pfalz *Eindeckers* in 1916.

Idflieg circulated captured Nieuport fighters to several German aircraft manufacturers and attributed the Nieuport's superiority over the *Eindeckers* to its sesquiplane (one and a half wing) wing. However, fighters with conventional biplane wings also out-performed the *Eindeckers*, a fact that apparently escaped *Idflieg*. Some manufacturers basically ignored the sesquiplane configuration, others copied the Nieuport, and others incorporated features of the sesquiplane wing in their designs.

Idflieg's advocacy of the Nieuport's sesquiplane design had a pernicious affect on the German aviation industry. The basic sesquiplane advantage – it afforded good downward visibility for the pilot – was more than offset by its main disadvantage – structural weakness. The problem was not static strength but flutter, a dynamic problem not then understood. A major mischief from *Idflieg's* shortsightedness was the waste of resources on this outdated technology. Another plague was the structural problems afflicting the Albatros D.III and D.V fighters that adopted the sesquiplane wing.

The only German copy of the Nieuport to see operational service was the SSW D.I, which was outdated by the time it reached the front, an outcome that should have been foreseen by *Idflieg*. Failing to learn from this debacle, *Idflieg* next jumped on the triplane configuration as a key to fighter performance, another significant error in judgment.

Above: This reproduction of a Nieuport 11 in Italian colors illustrates the Nieuport's sesquiplane, literally one and a half wing, configuration. The 'V' strut clearly indicates the lower wing had only one spar, the source of its structural problems. As light and relatively slow as the Nieuports were, they were subject to in-flight wing failures for this reason. The problems were even more severe for heavier, faster airplanes like the Albatros fighters.

Above: The SSW D.I was basically a copy of the Nieuport 17, but SSW used their own 110 hp Siemens-Halske Sh.I counter-rotary engine for better performance, which required a larger propeller. This early production model has no spinner.

Above Right: The spinner on this late-production SSW D.I gives it an aggressive look. The SSW D.I handled well but its performance was not competitive when it reached the front, which one would expect from copying an older design.

Above: The Kondor D.I appeared in 1918 and used the outdated sesquiplane design. Only the single prototype was built; the subsequent D.II used a two-spar lower wing.

Above: The Euler D.II was basically a copy of the Nieuport 11. It was out-dated when delivered and used only for training due to its limited performance.

Left: Possibly the worst problem stemming from the sesquiplane influence was its use on the Albatros D.III and follow-on D.V, which experienced serious structural problems from their single-spar lower wings. Here the attachment of the bottom of the 'V' strut indicates the single-spar lower wing. This Albatros D.III served with *Jasta* 19; *Lt.d.R.* Reinhold Oertelt is the pilot.

Below: The Pfalz D.III used a narrow-chord lower wing for good downward visibility, but used two spars for a strong wing cellule. Here is *Vzfw*. Rüdenberg of *Jasta* 10 in his D.III.

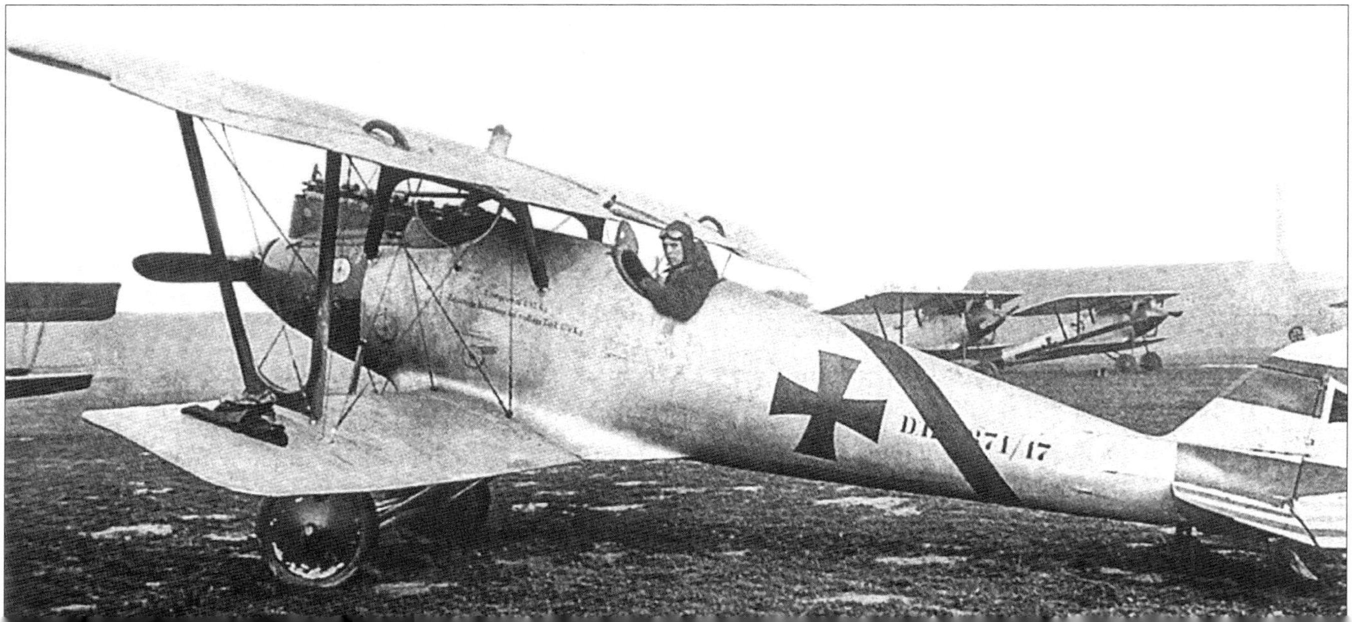

Why the Triplane Craze?

Above: This flying reproduction Fokker Triplane is in authentic colors for Manfred von Richthofen, the famous Red Baron. Prototype Fokker Triplanes arrived at the front in September 1917 and the Red Baron was killed in his triplane on 21 April 1918. This connection with the famous Red Baron makes the Fokker Triplane far more famous than it would be otherwise, and arguably makes it the most famous airplane of the Great War.

Introduction of the powerful Albatros fighters armed with two synchronized machine guns in the late summer of 1916, coupled with the formation of the first German fighter squadrons, gave the German Air Service superiority over the Western Front. Organization of fighter units matched the moves pioneered by the French early in 1916 over Verdun and soon adopted by the RFC and RNAS, while the Albatros returned the technical advantage to Germany, who had lost it when the French Nieuport fighters and the British DH-2 appeared over the front to challenge the Fokker and Pfalz monoplane fighters.

As more Albatros fighters were delivered and more Jastas were formed, German air superiority gained momentum, peaking in April 1917 with a sound tactical defeat of the Allied air arms in a debacle named "Bloody April" by the hard-pressed British flying services.

During this period of German resurgence over the Western Front, two new Allied fighters were introduced in small numbers that presaged Allied recovery. One was the Spad 7 biplane. Powered by the innovative new Hispano-Suiza V-8, the Spad was strong, fast, and maneuverable. Its main limitation as a fighter was that it only carried one synchronized machine gun. Building up to a significant production rate was also a problem, with the new engine being the main problem.

While the French introduced the sturdy Spad

biplane, the British RNAS introduced a more distinctive fighter. For sound structural and aerodynamic reasons, most airplanes before and during the war were either monoplanes or biplanes. However, in early 1917 Sopwith introduced a triplane development of their agile Pup fighter. The Sopwith Triplane retained the single gun and general lines of the Pup, but had a more powerful, 110 hp rotary coupled with a new triplane wing cellule.

Apart from their single gun armament, both the Spad and Triplane out-performed the Albatros. However, neither was available in sufficient numbers to prevent the Albatros from achieving air superiority in late 1916 and early 1917. While both the Sopwith and Spad designs were good, the Spad 7 had more development potential and it and its two-gun development, the Spad 13, remained at the front until the armistice.

Apparently it was the Sopwith's triplane configuration, unusual for a fighter, that was the key to its charm. While a good fighter, the Spad was a biplane, the most common configuration in WWI, and its performance could be attributed to its innovative engine. On the other hand, the Sopwith Triplane was the first triplane fighter to appear over the front, and its rotary engine was a known quantity. Regardless, the Sopwith Triplane caught the imagination of Idflieg, who imagined that its triplane configuration was the secret to its success.

Despite the overall German superiority during the Spring of 1917, the Sopwith Triplane caused Idflieg to think that perhaps the triplane had some inherent superiority to the biplane. To the contrary, the biplane configuration is inherently a better compromise between strength, weight, and aerodynamic drag than the triplane, which has too much drag.

However, the light, agile Sopwith Triplane had hit a lucky peak of triplane potential, and the result was the triplane craze, an affliction confined almost entirely to Germany and Austria. German authorities, followed by the Austrians, requested their manufacturers to develop triplane fighters, resulting in a plethora of new designs that never reached the Front and wasted a great deal of engineering resources in the process. Only two German triplane fighters reached production. One was the Pfalz Dr.I (Dr for *Dreidecker*, or triplane); only ten were built and while there are photographs of it at *Jasta* 73, its appearance in combat is not confirmed. The other was the Fokker Triplane, made famous as the airplane the Red Baron was flying when he was killed.

In many ways the Fokker Triplane was a paradox. It came at a time when the Albatros, Roland, and Pfalz fighters were inferior to their Allied opponents, so was welcomed warmly by the German pilots. Initially beset with some structural failures due to poor quality control during production, the problems were soon rectified and it became a favorite mount of German aces despite being slow, its main limitation. On the positive side, it had exceptional climb and maneuverability ideally suited to close-in, high-G maneuvering combat, or dog-fighting as it was called, that was so characteristic of WWI combat over the Western Front. It had the standard two synchronized machine guns and was, once the quality control problems were solved, a sturdy airplane. Despite the claims of some critics, it was not a copy of the Sopwith, sharing nothing in common with the Sopwith's design other than its triplane configuration and use of a rotary engine. While it needed more speed to be a first-class fighter, at least it out-performed its opponents in climb and

Left: The Sopwith Pup biplane fighter was powered by an 80 hp rotary and had one synchronized machine gun. Its excellent maneuverability and handling qualities made it a pilot's airplane and helped it hold its own against the two-gun Albatros fighters powered by a 160 hp Mercedes, but it was normally on defense against the poweful Albatros.

Allied Competitors

Nieuport 11 N.1256 of *Sgt.* Raoul Lufbery, N.124, June 1916

Sopwith Pup N6205 *Betty* flown by FSL Joseph S.T. Fall, No.3(N) Sqdn, RNAS, April–May 1917. Fall scored 3 of his 36 victories flying N6205.

Sopwith Triplane N533 *Black Maria* of Flt. Lt. Raymond Collishaw, No.10(N) Sqdn., RNAS, July 1917

Austin Osprey X15, Prototype February 1918; Britain did not abandon triplane fighters until well into 1918.

Above & Below: The cause of all the trouble, the Sopwith Triplane. Clearly related to the preceding Pup, the Triplane had an elegant triplane wing cellule of high aspect ratio with enough gap between the wings to minimize airflow interference between them. The single fixed gun remained, but the engine was now a 130 hp Clerget, an increase of 62.5% over the power enjoyed by the Pup. The Triplane had clean lines but was not as robust as the later Sopwith Camel biplane. The middle wings were attached to the cabane struts instead of directly to the fuselage, creating additional drag from vortices at the wing roots as well as the wing tips. Ailerons were fitted to all wings for a good roll rate.

maneuverability, whereas the Albatros was now inferior in all respects.

Eventually sanity prevailed, and designers abandoned the triplane configuration due to its higher drag. The Sopwith Triplane was replaced by the biplane Camel, which did not really outperform it but was much sturdier and had two guns. The Fokker Triplane was developed into the faster Fokker D.VI biplane. But the D.VII, powered by a water-cooled Mercedes or BMW, was a more potent fighter with higher ceiling andbetter speed and climb at altitude, and D.VI production was limited.

Above: Restored Spad 7 held in the National Museum of the USAF. The Spad 7 was a great success and was flown in combat by France, Belgium, Britain, Italy, Russia, and the USA.

Below: An early Spad 7 with the initial 140 hp Hispano-Suzia V-8. Engine power was soon developed to 150 hp, then 180 hp, making the Spad 7 faster than the Albatros. Only one synchronized gun was fitted. Arriving at the front about the same time as the Sopwith Triplane, the Spad did not make the same emotional impression on *Idflieg*, perhaps because it was a biplane. Regardless, the robust Spad airframe had more development potential than the Sopwith Triplane and the more powerful, two-gun Spad 13 became the standard French and American fighter until the end of the war.

Why a Triplane?

Airplane designs are challenging compromises between aerodynamic form for speed, strength to withstand the rigors of flight, and light weight to carry a useful load and still be able to fly. Most WWI designs were biplanes, with a number of monoplanes, but triplanes were rare. So when might a triplane configuration be a reasonable choice?

From the standpoint of aerodynamic drag, more wings mean more drag and lower speed, which is especially important for combat aircraft. Lower drag for greater speed seems to make the monoplane the preferred configuration. Yet most WWI airplanes were biplanes, not monoplanes. The reason is the wings must be strong enough to support several times the aircraft's weight to withstand maneuvering loads, and it is much easier and lighter to make a biplane strong enough than to make a monoplane strong enough. With the unsophisticated structural designs and materials of the time, making a strong monoplane either required a structure that was much too heavy or extensive bracing from wires or struts. The bracing wires themselves were heavy and created a lot of drag, essentially nullifying the aerodynamic advantage of the monoplane. All things considered, the biplane was generally a better choice than a monoplane because it could be made stronger and lighter. And due to the extensive bracing required for most monoplanes, the biplane's drag that was almost as low.

The advantage of a triplane configuration is the ability to concentrate a lot of wing area in a relatively small wing span, making a triplane more compact than a biplane or monoplane that could carry the same load. This is especially advantageous for bombers that need to carry heavy bomb loads, and there were a number of triplane bomber designs during WWI. The triplane advantage for a fighter is greater lift in a small size, which can enable triplane fighters to be more maneuverable and have better climb rates than biplanes or monoplanes.

The triplane disadvantages are more weight and drag; biplane fighters with the same engine are normally faster. The WWI emphasis on high-G close combat between fighters, or 'dogfighting', made good maneuverability essential for a fighter. However, even in WWI pilots recognized the need for speed, and many pilots thought speed was the most important fighter attribute. The priority for speed was a disadvantage for triplane fighters, and they soon disappeared from combat, superseded by faster biplanes.

Biplanes were the mainstream in WWI. However, in Germany, development of better structures and materials enabled strong, light monoplane designs, and by mid-1918 many German designers were building monoplanes to achieve even higher speeds than biplanes. Ironically, *Idflieg's* 1917 craze over triplanes gave way to monoplanes by late 1918.

Below: A huge Caproni triplane, tri-motor bomber in service with the RNAS. For a bomber, the triplane configuration offered a lot of wing area to lift a heavy bomb load. The Caproni Ca.4 below was one of the few triplane designs to see production and operational service. The profusion of struts and bracing wires created extensive drag, making it very slow.

Above & Below: The BFW N.I was yet another triplane bomber design. Intended as a night bomber, the triplane configuration enabled it to carry a heavy bomb load for a single engine. The attempt to reduce drag by the clean nose entry was spoiled by the flat radiator under the upper wing. Like the Caproni triplane bomber on the facing page, the BFW N.I was very slow, but unlike the Caproni it was not placed into production or service.

Triplane Fighter Prototypes

AEG Dr.I

According to AEG records, the AEG Dr.I fighter triplane was built to compare the performance with the D.I biplane, but it is no coincidence that the Dr.I arrived at the height of the triplane craze. Rolled out in October 1917, the Dr.I, like the D.I biplane, was powered by a 160 hp Mercedes D.III engine and armed with twin machine guns. The Dr.I demonstrated a slightly higher rate of climb than the D.I but, according to *Idflieg*, "the speed was drastically lower, precluding any consideration of front-line service." Only one prototype was built.

AEG Dr.I Triplane Specifications

Engine:	195 hp Benz Bz.IIIb V-8	
Wing:	Span Upper	9.40 m
	Span Middle	9.10 m
	Span Lower	8.80 m
	Chord Upper	1.37 m
	Chord Middle	0.75 m
	Chord Lower	0.75 m
	Gap Upper	1.00 m
	Gap Lower	0.75 m
	Area	24 m^2
General:	Length (fuselage)	6.15 m
	Track	1.0 m
	Empty Weight	710 kg
	Loaded Weight	970 kg
Maximum Speed:		200 km/h
Climb:	1000m	2.5 min

Above & Below: The AEG Dr.I was a triplane version of the AEG D.I. The additional lift gave it a slightly better rate of climb, but the additional drag dropped the speed significantly, so production was out of the question. Ailerons on the top and bottom wings were connected by an actuating strut. The initial version above has single-bay bracing. Later the aircraft was modified to have 2-bay wings as shown below, perhaps as a testbed for the DJ.I triplane, which used a similar wing cellule.

Above & Below: The initial version of the AEG Dr.I, a triplane version of the AEG D.I biplane fighter. The block radiators on either side of the fuselage created more drag than airfoil radiators, but the D.I biplane was a fast fighter prototype despite the radiators. The triplane wing cellule simply created too much drag to be competitive. (Both photos The Peter M. Bowers Collection/The Museum of Flight)

AEG DJ.I Triplane

It is not known when the idea of an armored, single-seat fighter to oppose Allied ground-attack aircraft was proposed, but surely it was another inspiration of *Oberst-Leutnant* Wilhelm Siegert, the commander of *Idflieg*, Siegert remained a steadfast proponent of specialized aircraft to fulfill specific tactical assignments throughout the war; he had no use for general-purpose aircraft. It was reported on 21 October 1917 that AEG was designing an armored fighter, known internally as the AEG PE (*Panzer-Einsitzer*) and may have been later designated DJ.I or DA.I. [In late 1918, two aircraft classifications appeared in various records, both signifying the armored single-seat fighter. They were DJ = D-*Infantrie* and DA = D-*Angriff* (attack). It is not known if these classifications were officially adopted.] The engine, fuel and pilot were protected by a steel shell forming an integral part of the airframe. The rear fuselage was of aluminum. The triplane wings had steel-tube spars and wooden ribs.

AEG DJ.I Triplane Specifications

Engine:	195 hp Benz Bz.IIIb V-8	
Wing:	Span Upper	11.20 m
	Span Middle	10.60 m
	Span Lower	10.00 m
	Chord Upper	1.475 m
	Chord Middle	0.825 m
	Chord Lower	0.825 m
	Gap	1.00 m
	Area	31.3 m^2
General:	Track	1.90 m
	Empty Weight	1,182 kg
	Loaded Weight	1,412 kg
Maximum Speed:		166 km/h
Climb:	1000m	5.8 min
	4000m	48 min

Preliminary trials of the AEG PE triplane, powered by a 195 hp Benz Bz.IIIb V-8 engine, were expected to be concluded in December 1917. From a report dated March 1918, we learn that the AEG

Above: The AEG DJ.I triplane was an armored fighter designed to defend against Allied ground-attack aircraft at low alritudes where armor against rifle and machine-gun fire would be very important for survivability. The DJ.I triplane was too slow, the common problem with triplanes, and a biplane version was then built and tested. The biplane was faster but was not placed in production, perhaps due to the armistice.

Above: The AEG PE (*Panzer Einsitzer*) single-seat fighter had an armor-plated nose protecting pilot, engine and fuel. Even the exhaust headers and side radiators are protected by armor plate. The fuselage behind the pilot was of constructed of aluminum. The twin-bay triplane rigging of the AEG PE was very unusual and reflects the great weight of the aircraft. The armament consisted of two synchronized machine guns (firing through the openings in the armored cowling) and four small bombs.

Panzer-Einsitzer had reached 4,000 meters (13,123 ft). in 48 minutes. The test pilot reported the triplane as easy to fly but not particularly fast. Work on the AEG DJ.I triplane was stopped in May 1918 because of "poor flight characteristics."

Disappointed by the low speed and mediocre flight characteristics of the AEG DJ.I triplane, AEG engineers brought out a biplane version, the AEG DJ.I fighter, also powered by the 195 hp Benz Bz.IIIb V-8 engine. Sharing the triplane's armored fuselage design, the DJ.I had a wireless wing cellule to minimize damage from ground fire. The construction began in March and was completed in May 1918, but according to *Idflieg* it was not until August that the flight tests began, possibly due to the shortage of the experimental Benz V-8 engine.

Below: The AEG DJ.I biplane used the fuselage from the preceding triplane. It had an armored nose protecting the pilot, fuel and engine. To reduce the danger from ground fire, the control cables were doubled and the wing cellule was wireless. At an empty weight of 1,185 kg (2,612 lb.), the DJ.I was almost three times heavier than the Fokker D.VIII fighter. The biplane version climbed to 1,000 m in 4 minutes and had a maximum speed of 180 km/h. Both climb and speed were superior to the triplane powered by the same engine.

Albatros Dr.I

The Albatros Dr.I was a triplane version of the D.V biplane fighter. The sole prototype was flown in the summer of 1917. Ailerons were fitted to all wings and armament was the standard two machine guns.

Albatros Dr.II Specifications		
Engine:	160 hp Mercedes D.III	
Wing:	Span	8.70 m
General:	Length	7.33 m

Above & Below: During the German triplane craze, no biplane fighter was safe from modification. Even the Albatros D.V was fitted with a triplane wing cellule for comparison, creating the Albatros Dr.I. The standard biplane D.V was faster, had better climb, and was more maneuverable due to the extra weight and drag of the triplane wings. (Both photos The Peter M. Bowers Collection/The Museum of Flight)

Albatros Dr.II

The Albatros Dr.II was a triplane version of the D.X biplane fighter prototype. The sole Dr.II prototype was flown in the spring of 1918. Ailerons were fitted to all wings and armament was the standard two machine guns.

Albatros Dr.II Specifications		
Engine:	195 hp Benz Bz.IIIbo	
Wing:	Span	10.00 m
	Area	26.6 m^2
General:	Length	6.18 m
	Empty Weight	676 kg
	Loaded Weight	915 kg

Above & Below: The second Albatros triplane fighter was the Dr.II, powered by an experimental V-8. The large, flat radiators created excessive drag and were surely fitted to expedite early test flights, with replacement as soon as practical by more streamlined radiators. Its counterpart, the D.X biplane, had better performance.

Brandenburg L 16

Although Brandenburg was a German company, the L 16 was built for the Austro-Hungarian *Luftfahrtruppen*. Early in the war some structural test failures of Brandenburg landplanes had so angered the German army that no more Brandenburg designs were accepted for evaluation by the army. This meant that Brandenburg had to sell to the German navy and the Austro-Hungarian Empire. The L 16 was the last Brandenburg landplane design; all subsequent Brandenburgs were seaplanes.

Brandenburg designed the L 16 after receiving technical data on the Sopwith Triplane. The stress calculations for the L 16 were submitted for approval on 13 August 1917. On 1 December 1917 Austro-Hungarian representatives inspected the completed L 16 and were struck by how large it was. The climb estimates were highly unlikely to have been achieved during flight testing, and after brief testing the L 16 was stored until the armistice.

Brandenburg L 16 Specifications

Engine:	200 hp Daimler	
Wing:	Span Upper	9.00 m
	Span Middle	9.00 m
	Span Lower	9.00 m
	Chord Upper	1.30 m
	Chord Middle	1.20 m
	Chord Lower	1.30 m
	Gap upper	1.20 m
	Gap lower	1.30 m
	Stagger	0.50 m
	Area	33.5 sq m
General:	Length	7.17 m
	Height	2.82 m
	Track	1.70 m
	Empty Weight	740 kg
	Loaded Weight	935 kg
Maximum Speed:		190 km/h
Climb (estimated):	1000m	1 min
	2000m	3 min

Left: The prototype L 16 ready for initial flight testing in Germany. The boxy, strut-mounted radiators were fitted to expedite flight tests. Engine was a 200 hp Daimler. The two successful triplane fighters, the Sopwith and Fokker, were small, light fighters that featured excellent climb and maneuverability. In contrast, the L 16, while streamlined, was large and heavy, negating the primary advantages of the triplane configuration.

Right: The prototype L 16 ready for initial flight testing in Germany. The original rudder had a rounded trailing edge. The L 16 was large for a fighter. (The Peter M. Bowers Collection/The Museum of Flight)

Above & Below: The Brandenburg L 16 during flight testing at Aspern in the Austro-Hungarian Empire. The rudder has been modified with a pointed trailing edge and a smaller radiator has been fitted in front of the cockpit, which hardly improved the field of view. One hesitates to imagine the effect on the pilot of a radiator leak from mechanical failure or combat damage. The broad interplane struts assisted streamlining but also obscured the pilot's view. (Both photos The Peter M. Bowers Collection/The Museum of Flight)

DFW T 34-II

Above: The DFW T 34-II triplane was developed in parallel with the T 43-I biplane fighter. Both were submitted to the First Fighter Competition but were excluded from flight trials due to a poor field of view for the pilot. (The Peter M. Bowers Collection/The Museum of Flight)

The DFW T 34-II, was a triplane fighter developed in parallel with the T 34-I biplane fighter prototype and shared everything but the wing cellule with its biplane counterpart.

Like its biplane sibling, the T 34-II was powered by a 160 hp Mercedes D.III engine and was armed with two machine-guns. Both the biplane and triplane were submitted to the First Fighter Competition but were excluded from flight testing due to poor field of view for the pilot. In addition, the triplane was condemned for "unsuitable design", although details are not available. No performance or dimensional data for either aircraft is available.

Left: The DFW T 34-I biplane was developed in parallel with the T 43-II triplane fighter. The similarity of the two designs is clear from this photo of the biplane taken at the First Fighter Competition. Both types were rejected due to the poor field of view afforded the pilot.

Above & Below: The DFW T 34-II was a compact triplane with many drag-creating bracing struts. The bottom wing was suspended below the fuselage to give more gap to prevent airflow interference between the wings. No performance data are available. (Both photos The Peter M. Bowers Collection/The Museum of Flight)

Fokker V4

Fokker created the only successful German triplane fighter. The production Fokker Dr.I was developed from the first Fokker triplane fighter prototype, the V4, which had cantilever wings with no bracing struts.

The V4 was an interesting design made possible by Fokker's exploration of relatively thick, cantilever wings using a wood box spar. The fuselage and tail were typical Fokker structures of welded steel tube covered by fabric. All three wings were of the same chord and the lower and middle wings were of the same span. Unbalanced ailerons were fitted to the upper wings only, which had longer span. For additional lift an airfoil was fitted around the axle of the undercarriage, a feature found on all subsequent production Fokker fighters of the war.

Flown for the first time in May 1917, the V4 exhibited good performance and flying qualities. Handling was improved by adding aerodynamic balances to the ailerons and elevators. The cantilever wings vibrated in flight, disconcerting the pilots, so interplane struts were added. These reduced wingtip vibration and enhanced wing torsional stiffness. Together with the horn-balanced ailerons, the interplane struts improved aileron effectiveness.

On 5 July a second V4 prototype was ordered. The first V4 was armed with two synchronized machine guns and shipped to Austria-Hungary in late August 1917.

Fokker V4 Specifications

Engine:	110 hp Le Rhône	
Wing:	Span Upper	6.21 m
	Chord Upper	1.20 m
General:	Length	5.75 m
	Height	2.95 m
	Empty Weight	346 kg
	Loaded Weight	528 kg
Maximum Speed:		200 km/h
Climb:	1000m	2 min
	2000m	4.5 min
	3000m	8 min
	4000m	13 min
	5000m	20 min

Above: The Fokker V4 triplane prototype had cantilever wings and originally had no interplane struts. Unbalanced ailerons were fitted to the top wing only.

Above & Below: The Fokker V4 triplane prototype, above without markings and below with markings. The leading edge of the horizontal stabilizer was rounded. The V4 introduced the airfoil over the undercarriage axle used by all subsequent Fokker WWI production fighters.

Fokker V5 & the Fokker Dr.I

The Fokker V5, ordered on 11 July 1917, was based on the final configuration of the V4. Powered by a 110 hp Le Rhône, copies of which were built as the Oberursel Ur.II, the V5 was the prototype for the production Fokker Dr.I

The main difference between the V5 and the earlier V4 was that the middle wing of the V5 was increased in span so it was intermediate in span between the upper and lower wings. Another visible change from the V4 was the shape of the horizontal stabilizer.

Fokker V5 Specifications

Engine:	110 hp Oberursel Ur.II	
Wing:	Span Upper	6.725 m
	Wing Area	16.0 m^2
General:	Length	5.75 m
	Height	2.725 m
	Empty Weight	375 kg
	Loaded Weight	571 kg
Maximum Speed:		200 km/h
Climb:	1000m	1 min
	2000m	4.5 min
	3000m	8.5 min
	4000m	14 min
	5000m	23 min
Armament:		2 mgs

Above: The Fokker V5 triplane was the prototype of the Dr.I production fighter. The middle wing was increased in span compared to the V4, giving different spans for all three wings.

Above & Below: The Fokker V5 triplane was the prototype of the production Fokker Dr.I. Based on the final configuration of the earlier V4, it had horn-balanced ailerons and interplane struts for excellent aileron effectiveness. The horn-balanced elevators differ from the final production configuration.

Above: The Fokker V5 triplane prototype is at right, with the earlier Fokker V3 biplane prototype at left.
Below: The Fokker V5 triplane prototype in flight.

Fokker V6

Developed in parallel with the rotary-powered V5, the Fokker V6 featured a 160 hp Mercedes D.III six-cylinder, water-cooled engine. Ordered on 7 July 1917, the V6 was a larger airplane than the V5, which was necessary to retain a similar wing loading despite the heavier engine. Wing span, chord, and area were all increased.

The greater chord compelled increased wing gap to avoid airflow interference between the wings. At first the bottom wing was mounted beneath the fuselage on short struts, but to reduce drag it was faired into the fuselage.

The heavy V6 lacked the good climb rate and sparkling maneuverability of the lighter, rotary-engine triplanes and development was dropped.

Fokker V6 Specifications

Engine:	160 hp Mercedes D.III	
Wing:	Span Upper	7.90 m
	Span Middle	6.60 m
	Span Lower	6.00 m
	Wing Area	22.5 m^2
General:	Length	5.75 m
	Height	2.725 m
	Empty Weight	637 kg
	Loaded Weight	880 kg
Climb:	1000m	3 min
	2000m	6.5 min
	3000m	12 min
	4000m	20 min
	5000m	34 min
Armament:		2 mgs

Above: The Fokker V6 triplane fighter prototype was powered by a 160 hp Mercedes D.III. Larger and much heavier than the V5, the V6 had poorer maneuverability and climb rate despite its greater power. Here it is seen after the fuselage had been deepened to fair in the lower wing for reduced drag.

Above: Originally the Fokker V6 triplane prototype had the lower wing attached to the bottom of the fuselage as shown here. Later, to reduce drag the lower wing was faired into the fuselage as shown here.

Below: The nose radiator on the V6 was new to Fokker designs and would reappear in later Fokker prototypes and the Fokker D.VII production fighter. The V6 is shown here after the lower wing was faired into the fuselage.

Fokker V7

Four V7 triplane prototypes were built from mid-August through late October 1917. All were based on the production Dr.I airframe but had more powerful rotary engines for improved performance together with longer fuselages to balance the heavier powerplants.

The V7S had the 11-cylinder, 160 hp Seimens-Halske Sh.III, the V7O had the 14-cylinder, twin-row, 160 hp Oberursel U.III, the V7G had the 9-cylinder, 160–200 hp Goebel Goe.III, and the V7R had the Austrian-built, 11-cylinder, 150 hp Steyr Le Rhône. The V7R was transferred to Austria as the MAG 90.03, where it crashed during testing.

The twin-row Oberursel was a heavy, outdated engine with problematic reliability, and gave little performance boost over the standard Dr.I at the cost of reduced maneuverability due to its weight. The Austrian-built Steyr was available in very limited numbers that were sufficient only for Austrian needs. Unfortunately, neither of the two more powerful rotaries, the Seimens-Halske Sh.III and Goebel Goe.III, were sufficiently developed to put into mass production. Furthermore, the required lubricant for rotary engines was in extremely short supply in blockaded Germany, so the V7 prototypes were not developed into production aircraft.

By this time it was becoming clear that the triplane configuration had too much weight and drag and Fokker turned his attention to biplane and monoplane fighters.

Fokker V7R (MAG 90.03) Specifications

Engine:	150 hp Steyr-Le Rhône	
Wing:	Span Upper	7.20 m
	Span Middle	6.22 m
	Span Lower	5.73 m
	Wing Area	17.5 m²
General:	Length	6.40 m
	Height	2.95 m
	Empty Weight	376 kg
	Loaded Weight	571 kg
Climb:	1000m	1.5 min
	2000m	4 min
	3000m	7.3 min
	4000m	11.5 min

Above Right: The V7R, powered by a 150 hp Steyr-built Le Rhône after transfer to Austria and redesignation as the MAG 90.03; the painter has not yet been informed of the serial number.

Right: The V7S powered by a 160 hp Siemens-Halske Sh.III engine under test at Adlershof. The large, four-blade propeller was needed to absorb the power of the Sh.III at low propeller RPM, and the tall undercarriage was needed to clear the propeller. No undercarriage airfoil was fitted. An enlarged cowling was needed for the Sh.III.

Pfalz D.III Triplane Conversion

To quickly produce a triplane fighter, Pfalz converted a standard Pfalz D.III to a triplane. The aircraft demonstrated slower speed and climb than a standard D.III, with poor flight characteristics. Interestingly, the same aircraft was flown as a biplane, with the middle wings removed, to determine the difference in performance between the biplane and triplane configurations. As a biplane the climb rate was better than as a triplane. The additional weight and drag of the additional wing plus airflow interference from the closely-spaced wings contributed to this poor performance.

Above & Below: In addition to the Pfalz Dr.I and Dr.II designed from the beginning as triplanes, Pfalz also converted a standard D.III to a triplane. Like the Albatros triplane conversion, the standard biplane was faster, more maneuverable, and had better climb rate due to its lower weight and drag.

Above: Like the standard Pfalz D.III, the Pfalz D.III triplane conversion was powered by a 160 hp Mercedes D.III. No designation for this aircraft is known, and no performance data or specifications have been found.

Below: Comparison of the Pfalz triplane conversion above with a standard Pfalz D.III biplane, 1405/17 below, shows the direct way Pfalz added the middle wing. To alleviate airflow interference the upper wing had to be raised, and this together with the new center wing undoubtedly restricted the pilot's field of view as well as reducing performance.

Pfalz Dr.I

The first Pfalz fighter designed from the beginning as a triplane was the Dr.I. The attractive Dr.I, powered by the then experimental 160 hp Seimens-Halske Sh.III counter-rotary engine, was well streamlined and had few bracing wires. During flight trials in August 1917 the Pfalz Dr.I demonstrated the fastest climb rate of any German WWI aircraft. It reached 5,000 meters in 11.5 minutes compared to the 35 minutes required by the Albatros D.V, then the standard German fighter. This stunning result generated great interest in introducing the Dr.I into combat in quantity. However, the Sh.III engine was too unreliable at the time, so production was limited to an evaluation batch of 10 Pfalz Dr.Is which were produced and sent to the front in April 1918. Although the aircraft had flown months before, engine unreliability prevented earlier deliveries. The

Pfalz Dr.I Specifications

Engine:	160 hp Siemens & Halske Sh.III	
Wing:	Span Upper	8.55 m
	Span Middle	8.10 m
	Span Lower	7.82 m
	Chord Upper	1.10 m
	Chord Middle	0.50 m
	Chord Lower	0.70 m
	Gap upper	0.78 m
	Gap lower	0.75 m
	Stagger	0.70 m
	Area	17 m^2
General:	Length	5.50 m
	Height	2.70 m
	Empty Weight	510 kg
	Loaded Weight	705 kg
Climb:	1000m	1.5 min
	2000m	3.0 min
	3000m	5.5 min
	4000m	8.5 min
	5000m	11.5 min
	6000m	14.5 min

Left: Pfalz Dr.I prototype on the Pfalz airfield. The 160 hp Seimens-Halske Sh.III counter-rotary engine required lots of cooling air.

Below: A large propeller was needed for best efficiency with the slow-turning counter-rotary engine. The combination gave the Pfalz Dr.I the best climb rate of any WWI German airplane.

Above & Below: Two views of Dr.I 3050/17, the second prototype, showing its armament. The Dr.I aircraft all were finished in the then-standard Pfalz silver-gray with standard iron crosses without white borders.

Pfalz Dr.I proved less maneuverable than the Fokker Dr.I and was too slow compared to contemporary biplane fighters. Its limited speed and engine development problems prevented mass production.

Above: In mid-December, 1917, *Rittmeister* Manfred von Richthofen had an opportunity to fly Pfalz Dr.I 3050/17 prototype which had recently returned to the Pfalz factory after flight tests at Adlershof. Manfred von Richthofen is in the middle of the group in a flying suit.

Above: This front view of Pfalz Dr.I 3050/17 emphasizes its clean lines. The large propeller, made by the Axial Propellerfabrik in Berlin, and many cowling cooling slots are shown clearly.

Pfalz Dr.II & Dr.IIa

Encouraged to develop the triplane configuration with a combat-ready engine, Pfalz followed the Dr.I with the slightly smaller Dr.II fitted with the proven 110 hp Oberursel Ur.II and the Dr.IIa fitted with the 110 hp Seimens-Halske Sh.I, which was more reliable than its more powerful development. Using low-powered engines, neither the Dr.II nor Dr.IIa had sufficient performance to merit production.

Although the Pfalz Dr.I, and its smaller derivatives the Dr.II and Dr.IIa, were not operational successes, the basic Dr.I design was developed into the single-bay Pfalz D.VII and two-bay D.VIII rotary-powered biplane fighters, both of which were placed into production. The stronger D.VIII was produced first and saw combat; the single-bay D.VII required more development and went into production too late to see combat.

Pfalz Dr.II & Dr.IIa Specifications

Engine:	Dr.II - 110 hp Oberursel Ur.II	
	Dr.IIa - 110 hp Siemens-Halske Sh.I	
Wing:	Span Upper	7.20 m
	Chord Upper	1.20 m
	Chord Middle	0.50 m
	Chord Lower	0.80 m
	Gap Upper	0.82 m
	Gap Lower	0.69 m
	Stagger	0.73 m
General:	Length	5.95 m
	Dr.II: Empty Weight	400 kg
	Loaded Weight	596 kg
	Dr.IIa: Empty Weight	395 kg
	Loaded Weight	568 kg
Climb: Dr.IIa:	1000m	3 min
	2000m	6.5 min
	3000m	10.2 min
	4000m	14.4 min
	5000m	19.4 min
	6000m	28.9 min

Above: Mechanics work on the unarmed Pfalz Dr.II powered by the 110 hp Oberursel Ur.II. One of the Pfalz Dr.I prototypes sits along side. The larger upper gap of the Dr.II compared to the Dr.I, intended to reduce airflow interference between the upper and middle wings, is clearly shown. This photograph was taken at Adlershof in March 1918.

Sablatnig SF4 Triplane

Above: The SF4 triplane carried Marine #901 and had ailerons on all wings. Little information is available on the triplane SF4, but it was not selected for production.

Designed as a single-seat floatplane fighter, the SF4 was built in both biplane and triplane versions. Both were powered by a 150 hp Benz Bz.III and carried one fixed, forward-firing Spandau machine gun.

The biplane, Marine #900, was tested first, and while speed was competitive, it had the lowest climb rate of all the single-seat floatplane fighter competitors. Worse, its maneuverability was poor due to its large wingspan and, despite its multitude of bracing wires, its structure was insufficiently robust; wing vibration was excessive in even a shallow dive. To improve the climb rate a triplane version, Marine #901, was built; like the biplane it was not competitive.

Sablatnig SF4Dr Specifications

Engine:	150 hp Benz Bz.III	
Wing:	Span	9.25 m
	Area	28.38 m^2
General:	Length	8.33 m

Company founder Josef Sablatnig had a doctorate in mechanical engineering and was a well-known pioneer pilot, so it is especially disappointing that he was unable to design an airframe that was at once light, strong, and streamlined. Although nose entry was streamlined, the wing structure created a lot of drag due to extensive bracing wires, and despite that the wing structure was weak.

Above: The SF4 triplane, Marine #901; the middle wings were attached to the cabane struts at their inboard ends rather than being attached to the fuselage. This enabled a larger gap between wings, reducing airflow interference. However, each middle wing now created a wingtip vortex at both ends, significantly increasing the induced drag and reducing speed and climb. The tail surfaces of the triplane version differed noticeably from the biplane version.

Right: The sole Sablatnig SF4 biplane fighter was Marine Number 900. Its climb and maneuverability were poor, so a triplane version was built. The triplane was not competitive with other seaplane fighters and was not produced in quantity.

SSW DDr.I Triplane
Fighter Prototype

AEG Dr.I Triplane
Fighter Prototype

Pfalz D.III Triplane Conversion
Fighter Prototype

Siemens-Schuckert DDr.I

The bizarre SSW DDr.I employed centerline thrust from two rotary engines but was not stable enough and development was abandoned.

Seimens-Schuckert DDr.I Specifications		
Engines:	2 x Dr.II - 110 hp Oberursel Ur.II	
Wing:	Span, Upper	7.20 m
	Wing Area	18.1 m^2
General:	Length	5.3 m
	Empty Weight	510 kg
	Loaded Weight	695 kg
Climb: Dr.IIa:	4,700m	20.6 min

Left & Below: The SSW DDr.I used two 110 hp Sh.I engines, one mounted in front of the nacelle and the other mounted in the rear of the nacelle. Two synchronized guns were fitted. The eccentric design crashed on its first test flight before performance data could be recorded.

Right: The SSW DDr.I. after its first test flight. Despite this setback, it was rebuilt, the wings being enlarged, and testing continued. It seems likely the drag from the lattice tail would have combined with that of the triplane configuration to make the airplane too slow, the congenital triplane problem.

**Albatros Dr.I Triplane
Fighter Prototype**

**Albatros Dr.II Triplane
Fighter Prototype**

**Schütte-Lanz Dr.I Triplane
Fighter Prototype**

LFG Roland D.IV / Dr.I

Originally designated the D.IV, Roland's only triplane fighter was re-designated the Dr.I when *Idflieg* promulgated the 'Dr' class designation for *Dreidecker* (triplane). This was the first Roland fighter to use the *klinkerrumpf* fuselage construction (clinker-built or lapstrake like a boat hull). Power was the 160 hp Mercedes D.III and armament was to be the standard two synchronized guns. The prototype crashed at the end of September 1917 and was rebuilt using a Roland D.VI fuselage, but further testing led to the design being abandoned.

Roland D.IV / Dr.I Specifications

Engine:	160 hp Mercedes D.III	
Wing:	Span	9.45 m
General:	Length	7.32 m
	Empty Weight	930 kg
	Loaded Weight	1208 kg
Maximum Speed:		155 km/h
Service Ceiling:		6,000m

Above & Below: The Roland D.IV / Dr.I used *klinkerrumpf* fuselage construction to save construction time.

Schütte-Lanz Dr.I

Although Luftfahrzeugbau Schütte-Lanz primarily built airships, it also built a number of prototype warplanes, none of which reached production. Encouraged to develop the triplane configuration with a production engine, Schütte-Lanz followed its Mercedes-powered D.III biplane design with a triplane development of that aircraft powered by the same engine and intended for the same armament of two synchonrized machine guns. The Dr.I used the entire fuselage and tail of the D.III mated to a triplane cellule that situated the upper two wings vertically with the lower wing staggered to the rear. This unusual wing arrangement may have been chosen to improve the pilot's field of view. The Dr.I was an unsuccessful participant in the Second Fighter Competition (27 May–28 June 1918). .

Schütte-Lanz Dr.I Specifications

Engine:	160 hp Mercedes D.III	
Wing:	Span	8.00 m
General:	Length	6.26 m
	Loaded Weight	900 kg

Above: The Schütte-Lanz Dr.I triplane fighter.

Left: The Schütte-Lanz D.III biplane fighter prototype was powered by a 160 hp Mercedes and carried the standard German fighter armament of two machine guns. It participated in the First Fighter Competition (20 Jan. to 12 Feb. 1918).

Below: The Schütte-Lanz Dr.I triplane fighter prototype was out-performed by the D.III biplane from which it was derived.

Above & Below: The Schütte-Lanz Dr.I, their entry into the triplane craze, was derived from their D.III biplane by substitution of a triplane wing cellule for the original biplane wings. The Dr.I, which participated in the Second Fighter Competition, was out-performed by its biplane predecessor, the Schütte-Lanz D.III.

Euler Triplane Fighters

In 1912 August Euler, owner of the Euler-Werke, patented a fixed, forward-firing machine gun aimed by pointing the aircraft at the target. Demonstrated to military authorities on 10 May 1912 on an Euler two-seat pusher, nothing further was done. But Euler persisted in fighter design, first with a single-seat pusher biplane, then with a two-seat pusher biplane, and then with the Euler D.I and D.II, which were copies of the Nieuport 11. A few Euler D.I fighters were evaluated at the front, but the D.I and D.II were used almost exclusively as trainers.

Euler *Dreidecker* Type 2

The first Euler triplane fighter design was the Euler *Dreidecker* Type 2, apparently completed in July 1917. This portly, inelegant triplane was powered by a 160 hp Oberursel U.III two-row rotary. Interestingly, the interplane struts had a 'Y' profile from the side, with one strut between the middle and bottom wing and two struts between the top and middle wings. Poor performance led to rebuilding the aircraft as a biplane. The biplane version was tested until April 1918. No specifications are available.

Euler *Dreidecker* Type 3

The next Euler triplane fighter design was the Euler *Dreidecker* Type 3, a completely new design powered by a 160 hp Mercedes D.III that first flew in November 1917. Like the earlier The first Euler triplane fighter design was the Euler *Dreidecker* Type 2 its performance was not impressive and it was rebuilt into a biplane for further testing.

Above: The first Euler triplane was this prototype trainer. Flown in August 1916, it was powered by a 220 hp Mercedes D.IV and had side-by-side seating for pupil and instructor. The engine was in demand for operational aircraft and only one was built. Span was 10.0 m, length 8.0 m, height 3.1 m, wing area 37.5 m^2.

Above: The Euler *Dreidecker* Type 2 was powered by a 160 hp Oberursel U.III two-row rotary.

Right: The Euler *Dreidecker* Type 3 was powered by a 160 hp Mercedes D.III engine.

Right: The Euler *Dreidecker* Type 3 was a completely different design from the earlier Euler *Dreidecker* Type 2. However, it too remained a single prototype.

Left: The Euler *Dreidecker* Type 3 gave disappointing performance and was rebuilt into a biplane for further testing.

Euler *Dreidecker* Type 4

The Euler *Dreidecker* Type 4, powered by the 180 hp Goebel Goe.III rotary engine, was the most successful Euler triplane prototype. It was demonstrated during the Second Fighter Competition because it climbed to 5,000 meters in 13.5 minutes. However, other than its climb rate the service evaluation pilots did not like the design and it was returned to Euler. No specifications are available.

Euler *Dreidecker* Type 5

The Euler *Dreidecker* Type 5, a triplane development of the Euler Quadraplane, was initially powered by the 100 hp Oberursel U.I rotary engine. It was re-engined with the 110 hp Siemens-Halske Sh.I and the wing gap was enlarged for better aerodynamic efficiency. Apparently the Type 2 biplane fighter was developed from this design. No specifications are available.

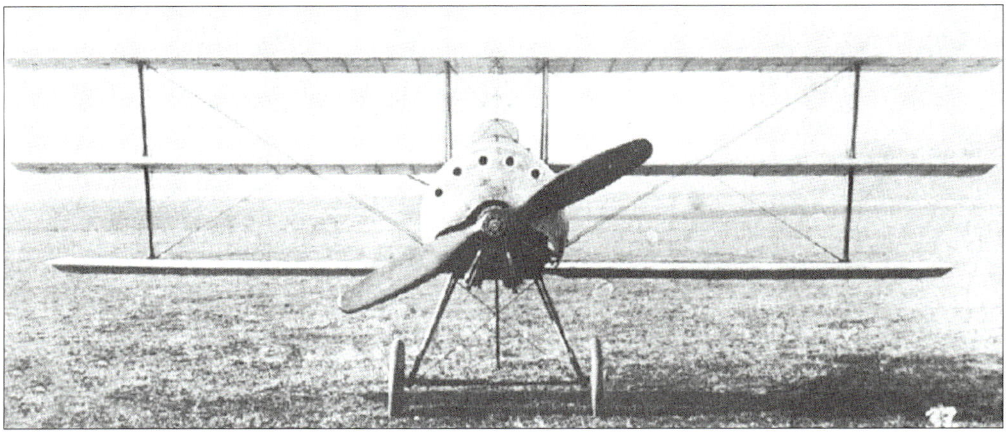

Top, Above, & Right: The Euler *Dreidecker* Type 4 was powered by the 180 hp Goebel Goe.III rotary engine and climbed to 5,000 m in 13.5 minutes.

Above: Euler *Dreidecker* Type 5 in its initial configuration.

Right: Euler *Dreidecker* Type 5 after re-engining with the Siemens-Halske Sh.I and the wing gap had been increased to reduce airflow interference between the wings.

Multiplane Madness

Triplanes were not the worst of it. Some designers reasoned that if more wings were better, than why stop at a triplane? Why not a quadraplane? Or even five wings? In retrospect this seems extreme or silly depending on your viewpoint, but multiplanes were tried, and not just by marginal designers. The sober and successful Friedrichshafen company designed and flew a quadraplane fighter, and the famous Anthony Fokker even tried five wings in his relentless search for superior performance!

Euler Quadraplane Fighter

Above & Below: Not to be outdone in the number of wings, Euler built a prototype quadraplane fighter powered by a 100 hp Oberursel U.I rotary engine. Wing area was 46.175 m^2, empty weight was 883.5 kg, and loaded weight was 1,383.5 kg, very heavy for the engine power. it appears to be derived from the Euler Dr.III.

Above & Left: Two more views of the prototype Euler Quadraplane. The control system was interesting in that the entire top wing rotated about the single spar to act as ailerons. This technically made it a triplane with full-span ailerons instead of a quadraplane. Powered by a 100 hp Oberursel U.I, it started flight testing in December 1917. This under-powered combination demonstrated poor performance which precluded further interest in the aircraft. Despite that, a second prototype with 110 hp Goebel Goe. II rotary engine was built. Empty weight was 883.5 kg, loaded weight was 1,386 kg, and wing area was 46.17 m^2. It was developed into the Euler *Dreidecker* Type 5 triplane, which in turn was developed into the Euler Type 2 biplane fighter prototype.

Friedrichshafen FF54

The FF54 was Friedrichshafen's response to the Triplane Craze at *Idflieg*. Designed by *Dipl.-Eng.* van Gries, it was a single-seat fighter quadraplane powered by the typical 160 hp Mercedes D.III engine and intended to be armed with the two machine guns standard for German fighters.

The narrow-chord middle wings were mounted to the fuselage with a narrow gap, and the interplane struts between the lower three wings were of broad, airfoil section to reduce drag. The conventional interplane struts between the top wing and upper middle wing were doubtless used to avoid obscuring the pilot's field of view. Flight testing was planned for late November 1917 but the exact date is unknown.

Flight testing soon showed the quadraplane configuration was not successful, and in April 1918 the FF54 was rebuilt as a triplane by removing the lower middle wing and revising the vertical tail surfaces. Renewed flight testing of the FF54 in triplane form began in May 1918, but after a crash in September 1918 the program was terminated. Friedrichshafen was successful designing bombers and floatplanes, but not single-engine landplanes.

Above: The Friedrichshafen FF54 is shown here in its original quadraplane form. No doubt the narrow gap between the pair of middle wings created airflow interference between them and contributed to the aircraft's failure. The lower middle wing was then removed and the vertical tail redesigned for renewed flight testing as a triplane. However, the FF54 was not successful as a triplane either. No specificiations or performance information has survived.

Right: The FF54 after conversion to a triplane. The revised rudder design is clearly shown.

Naglo D.II

Built by the Naglo Werft of Pichelsdorf near Berlin, apparently the Naglo D.II quadraplane was designed on the basis of 'more is better'. The retrogressive quadraplane wing cellule was added to a fuselage derived from the Albatros D.V. The Naglo D.II was evaluated in conjunction with the Second Fighter Competition and debriefing notes indicate it was to be modified and re-evaluated. Powered by a Mercedes D.III engine, it was intended to carry the standard German fighter armament of two machine guns. Official type testing was done 24 May 1918, and evaluation pilots praised the contruction and workmanship but wanted improved flying qualities. Designed by *Ing*. Gnädig, who was employed by Albatros at the time, its appearance resulted in his termination for conflict of interest and the prototype was rejected at the request of Albatros. The pylon for attaching the lowest wing was built into the fuselage. No performance data has survived.

Naglo D.II Specifications		
Engine:	160 hp Mercedes D.III	
Wing:	Span	9.00 m
	Area	22.40 m^2
General:	Empty Weight	914 kg

Above & Below: The Naglo D.II fighter prototype, either serial 1161/18 or 1165/18, was evaluated in conjunction with the Second Fighter Competition. The bottom wing was mounted on a keel built into the fuselage.

Fokker V8

Perhaps the worst example of a multiplane has been saved for last. What can one say about the Fokker V8 prototype fighter that was built on Anthony Fokker's instructions? Fokker had a reputation for an intuitive feel for aircraft design, but the V8 makes it clear that his intuition was developed through an arduous process of trial and error, and the V8 was clearly one of the errors.

The V8 used a combination of V6 triplane fuselage and tail combined with wings from a standard Fokker triplane. A full cellule of three wings was used at the front, and the upper two wings from another set of triplane wings was used at the center of the aircraft. Power was a 160 hp Mercedes D.III.

It should have been apparent this combination would result in a center of lift too far aft, producing unpleasant and perhaps dangerous flight characteristics and requiring a large download on the tail to balance the aircraft in level flight, thus reducing performance. If that was not clear from looking at the design, it was quickly made obvious during two brief test hops, after which the V8 was mercifully scrapped.

Fokker V8 Specifications

Engine:	160 hp Mercedes D.III	
Wing:	Span Upper	7.70 m
General:	Length	6.80 m
	Height	2.80 m
Armament:		2 mgs

Above: The bizarre Fokker V8 prototype used a Triplane wing cellule in front and the upper two wings of a Triplane aft. Two short test flights proved the V8 had dangerous handling qualities and was immediately abandoned. WWI aircraft development was mostly emperical, not theoretical, and the fact Fokker tried this five-wing monstrosity shows that the secret to his success was a constant series of experiments — many of which did not work! Proof that even the best designers can get carried away, the Fokker V8 exceeded the limits of good judgment. The old aviation axiom that, 'if it looks right it will fly right', is usually, although not always, true. However, there is a corrollary; 'if it looks wrong, it will not fly right', and the Fokker V8 is a prime example. Learning his lesson, Fokker recovered from this debacle and went on to design the exceptional D.VII biplane and innovative E.V/D.VIII monoplane fighters.

Austro-Hungarian Contagion

Germany's major World War One ally was the Austro-Hungarian Empire. While having its own aviation industry and a number of excellent aeronautical scientists and engineers, the Dual Monarchy was much less industrialized than Germany and produced about one tenth the number of aircraft.

As Germany's junior partner, the Austro-Hungarian Empire's aviation industry and *Luftfahrtruppe* were heavily influenced by their German allies. Thus, when Germany went triplane crazy for a time, the Austro-Hungarian aviation industry also joined in, if not as whole-heartedly, and produced several triplane fighter prototypes.

Information on the Austro-Hungarian fighter prototypes is presented here. For scale drawings of these interesting aircraft, see *Austro-Hungarian Army Aircraft of World War One* by Peter M. Grosz, George Haddow, and Peter Schiemer published by Flying Machines Press in 1993.

Lloyd 40.15

The Lloyd 40.15 was an eccentric triplane fighter prototype powered by a185 hp Daimler (MAG) engine. Drawings but no photos exist of this unusual airplane that was limited to a single prototype.

Lloyd had a reputation for unusual design features and the 40.15 prototype was no exception. The wings were fully cantilevered and roll control was provided by rotating wingtips on the central wing instead of conventional ailerons. The lower wing was mounted well below the fuselage behind the undercarriage. Two synchronized machine guns were mounted near the cockpit.

Design work began in late summer 1917, at the height of the triplane craze, and the airplane was delivered in December. Unfortunately, no information is available regarding performance or flying qualities.

Lloyd 40.15 Specifications

Engine:	185 hp Daimler	
Wing:	Span Upper	7.60 m
	Span Middle	5.80 m
	Span Lower	4.20 m
	Chord Upper	1.64 m
	Chord Middle	1.27 m
	Chord Lower	1.00 m
	Gap Upper	1.34 m
	Gap Lower	0.80 m
	Area	22.2 m²
General:	Length	7.10 m
	Height	2.82 m
	Track	1.60 m
	Loaded Weight	900 kg

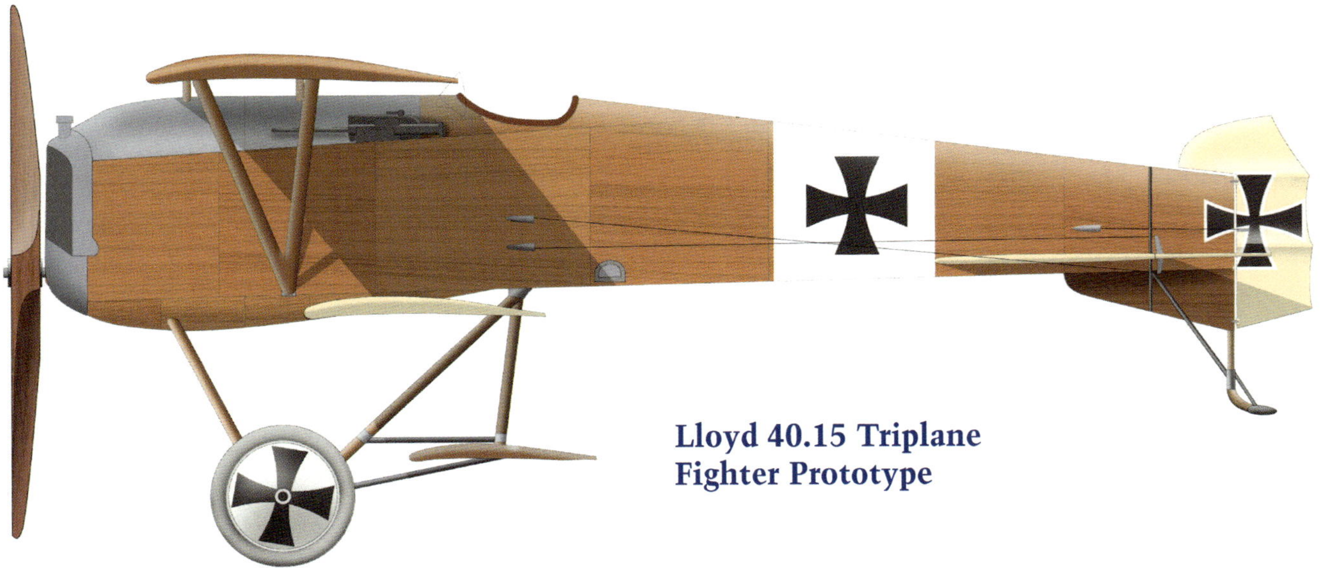

Lloyd 40.15 Triplane Fighter Prototype

Aviatik (Berg) 30.24

On 17 September 1917, at the height of the triplane craze, Aviatik submitted drawings for the 30.24 triplane fighter prototype based on the production Aviatik D.I biplane fighter. That same month Aviatik was given a contract for four experimental triplane fighters. All were completed but apparently only 30.24 was assembled and flown.

Unsurprisingly, flight tests in October 1917 revealed the 30.24 had lower performance than a similary-powered Aviatik D.I biplane. That should have been the end of the story, but 30.24 was returned to Aviatik for installation of a 200 hp Daimler and the standard twin-gun fighter armament. In late February 1918 flight tests of the modified airplane showed little performance improvement.

The 30.24 was then used for experiments with different radiators to improve the pilot's forward view for the production Aviatik D.I biplane. It existed until at least April 1920, when it was offered for sale to Czechoslovakia.

Aviatik 30.24 Specifications

Engine:	185 hp Daimler	
Wing:	Span Upper	7.23 m
	Span Middle	7.13 m
	Span Lower	7.00 m
	Gap Upper	0.85 m
	Gap Lower	0.85 m
	Area	22.73 m^2
General:	Length	6.25 m
	Height	2.75 m
	Empty Weight	730 kg
	Loaded Weight	900 kg
Maximum Speed:		174 km/h
Climb:	1000m	2 min 40 sec
	2000m	4 min 9 sec

Aviatik 30.24 was the only Austro-Hungarian triplane fighter deisgn that was adapted from an existing production biplane fighter.

Below: This side view of Aviatik 30.24 during flight testing at Aspern in October 1917 clearly reveals that it used an Aviatik D.I fuselage and tailplane modified to take a new triplane wing cellule. To achieve sufficient gap between the wings to minimize airflow interference, the lower wing was suspended below the fuselage by four struts. The camouflage scheme was one of many used on production D.I biplane fighters. As was generally the case with triplane conversions of biplane designs, the additional weight and drag of the third wing reduced speed and climb compared to the standard D.I biplane. Using an engine of greater power would not change the relative performance between the biplane and triplane configurations, so it is difficult to understand why 30.24 was re-engined.

Above: Front view of 30.24 shows the unusual four-blade propeller with angles of 70° and 110° between blades instead of the typical 90°. Wind tunnel experiments had shown this configuration to be aerodynamically more efficient.

Below: Rear view of the Aviatik 30.24 triplane fighter prototype. The cellon panels in the rear inboard sections of the middle wing to improve the pilot's downward field of view are clearly visible. Some of the rigging wires run through the center wing, and the ailerons, on the upper wing only, feature wash-out. The ailerons of many WWI airplanes featured washed out incidence to ensure the center of the wing stalled before the outer tips so the pilot would retain roll control during approaches to stall, which would help him avoid a spin.

Above: Aviatik 30.24 in front of the company hangar at Aspern. The cellon panels in the wingroots of the middle wing to improve the pilot's view downward are evident, as are the aerodynamically-balanced ailerons that were not used in other fighters designed by Berg. Behind 30.24 is Aviatik D.II 39.01, and Aviatik D.I 138.17 is in the background.

Above Right & Right: *Feldwebel* Joseph Kiss, an ace who scored 19 confirmed victories, is in the middle of this group. The modified 30.24, now with 200 hp Daimler and new, bulky radiator, provides a suitable backdrop for this group of aviators. The bulky new radiator definitely did not improve the aircraft's appearance.

Lohner Dr.I

The Lohner Dr.I, serial 111.04, was completed 23 June 1917, becoming one of the first triplane fighters flown by Germany and the Austro-Hungarian Empire. Designed by Leopold Bauer, its design and production were approved on 7 May 1917. Like many triplane fighter prototypes, it was a triplane development of an earlier biplane fighter, in this case the Lohner D.I 111.03 prototype. It was powered by a 185 hp Daimler engine. Flown during early July, it was dismantled on 8 July 1917 for small structural modifications. It returned to Aspern in August, and flight testing continued. Climb tests showed a lower climb rate than the contemporary Aviatik D.I biplane then entering operational service.

Test pilots reported the cockpit visibility was unsatisfactory and flight characteristics were totally unsatisfactory. Flight testing continued through October 1917, but the sole prototype was stored and finally written off on 11 October 1918.

Lohner Dr.I Specifications

Engine:	185 hp Daimler	
Wing:	Span Upper	8.80 m
	Span Middle	8.10 m
	Span Lower	8.80 m
	Chord Upper	1.00 m
	Chord Lower	1.00 m
	Area	24 m^2
General:	Length	6.35 m
	Height	3.0 m
	Empty Weight	693 kg
	Loaded Weight	946 kg
Maximum Speed:		178 km/h
Climb:	1000m	2 min 8 sec
	2000m	6 min 5 sec
	3000m	11 min 0 sec
	4000m	16 min 7 sec
	5000m	26 min 0 sec

Above: Lohner Dr.I 111.04 photographed before the serial number was applied. Flight testing revealed performance inferior to the Aviatik D.I biplane fighter already in production and squadron service, and test pilots regarded its flight characteristics as totally unsatisfactory.

Above & Below: Only one Lohner Dr.I prototype was built. The workmanship and finish were good, but the design itself was not satisfactory. The Dr.I was slightly faster and had a better climb rate than the Lohner D.I biplane with the same engine, which was unusual for a triplane development of an earlier biplane design. Regardless, poor flying qualities doomed the design. The the purpose for the unusual slanted interplane struts is not known.

Above & Below: Two more views of the sole Lohner Dr.I prototype showing the unusual slanted interplane struts. Poor flying qualities doomed the design and its performance was inferior to that of the Aviatik D.I biplane fighter already in service.

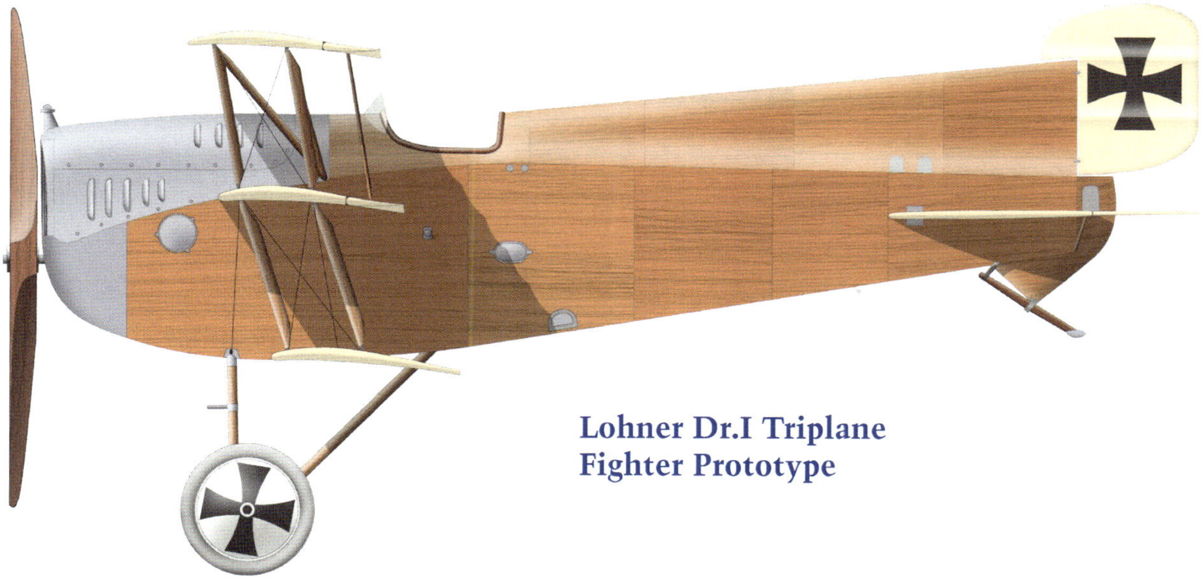

Lohner Dr.I Triplane
Fighter Prototype

Aviatik Triplane Fighter Prototype
30.24 in its Original Configuration

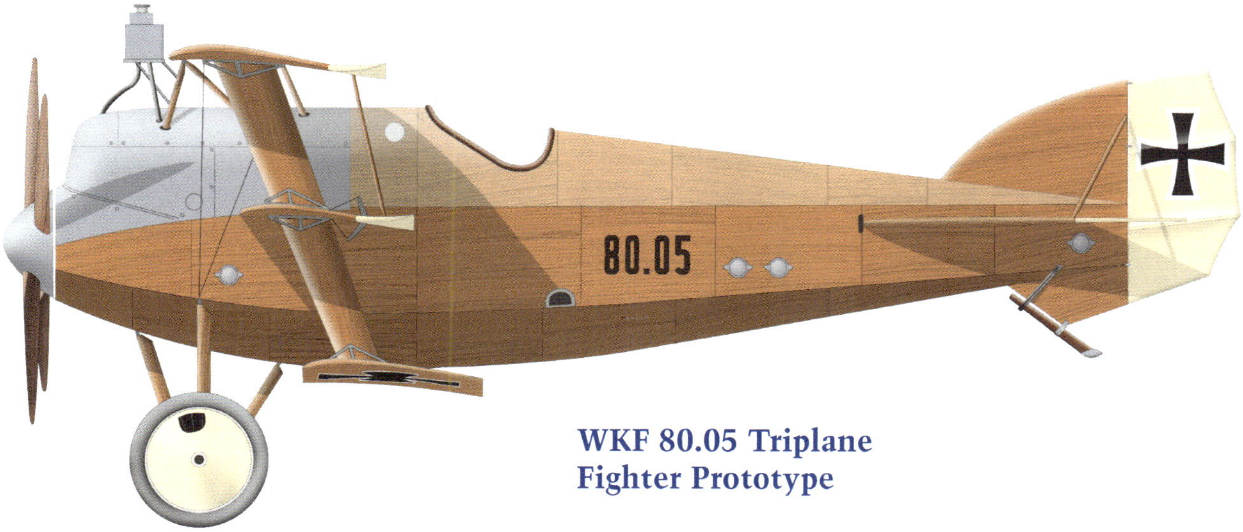

WKF 80.05 Triplane
Fighter Prototype

Oeffag 50.14

Construction of the Oeffag 50.14 prototype triplane fighter started in late 1917 when the triplane craze was at its peak. A 225 hp Daimler was installed in the airframe in April 1918.

Few details are available, but the two-piece upper wing was replaced by a single-piece wing during flight testing. Elevator area was also increased and the 50.14 was re-engined with a 200 hp Daimler. It appeared at the Austro-Hungarian Fighter Evaluation at Aspern in July 1918. However, both performance and maneuverability were mediocre and further development was cancelled.

Oeffag 50.14 Specifications

Engine:	225 hp Daimler	
Wing:	Span Upper	8.40 m
	Span Middle	8.40 m
	Span Lower	8.40 m
	Chord Upper	1.20 m
	Chord Lower	0.90 m
	Gap Upper	1.13 m
	Gap Lower	0.85 m
	Area	23.8 m^2
General:	Length	6.40 m
	Height	2.98 m
	Track	1.55 m
	Loaded Weight	970 kg

Above: The unarmed prototype Oeffag 50.14 triplane fighter in its original state with two-piece upper wing. The bulky, inelegant airplane had mediocre maneuverability. Despite having a powerful 225 hp Daimler, performance was also mediocre, so further development was cancelled.

Above: The view from the cockpit of the Oeffag 50.14 triplane fighter in its original state with two-piece upper wing.

Below: Lohner D.I 111.03 was the third prototype of the D.I biplane fighter and the immediate predecessor to the prototype Dr.I, 111.04. Neither the Lohner D.I nor Dr.I were successful.

WKF 80.05

The WKF 80.05 prototype triplane fighter was designed by Ing. Alfred Gassner, who had previously worked for Aviatik. Gassner started the design in September 1917 at the height of the triplane craze.

The fuselage had a hexagonal cross section typical of WKF designs. The wings and tailplane were covered in wood veneer instead of fabric. Ailerons were fitted to the top and middle wing, and the bottom wing was suspended below the fuselage to provide sufficient gap to prevent airflow interference between the wings.

Powered by a 200 hp Daimler, the 80.05 arrived at Aspern for testing in October or November 1917 for flight testing. Climb rate was less than the lighter, rotary-engine triplane fighters as might have been expected. To improve climb larger wings were fitted in March 1918, but further flight test results are not available. However, the 80.05 remained a single prototype.

WKF 80.05 Specifications (Final Configuration)

Engine:	200 hp Daimler	
Wing:	Span Upper	8.00 m
	Span Middle	7.80 m
	Span Lower	7.60 m
	Chord Upper	1.20 m
	Chord Middle	0.95 m
	Chord Lower	1.00 m
	Area	22.49 m^2
General:	Length	6.82 m
	Height	2.70 m
	Track	1.7 m
Climb:	3000m	11 min

In comparison, the WKF 80.06 biplane fighter, essentially a biplane version of the 80.05, was developed into the WKF D.I biplane fighter that was ordered into production and arrived just too late to see combat. Once again, a biplane proved superior to a similar triplane in performance.

Above: Initial version of the WKF 80.05 prototype triplane fighter at Aspern for flight testing in late 1917.

Above & Below: Two views of the sole WKF 80.05 triplane fighter prototype in its first configuration in the factory. The shiny surfaces of the veneer-covered wings, fuselage, and tail are cleary evident.

Above & Below: The WKF 80.05 triplane fighter prototype was surpassed in performance by the similar 80.06 biplane prototype, which was developed into the WKF D.I biplane fighter that entered production just too late for combat.

Above & Below: Two views of the WKF 80.05 triplane fighter prototype in its second, long-wingspan configuration. The four-blade propeller with 70°/110° angle between blades was typical of Austro-Hungarian late-war practice.

German Triplanes at War

Left: *Lt.* Werner Voss in Fokker F.I 103/17, a production prototype of the Fokker Dr.I, one of the iconic aircraft of WWI. This was partly due to the Red Baron being killed in his Fokker Triplane. Unfortunately, Voss met the same fate in a famous combat with No.56 Squadron, RFC. The Triplane had great climb and great maneuverability, but was slow. One innovation the Dr.I introduced was the airfoil over the axle between the wheels. The axle already created drag; the airfoil contributed lift as well. The undercarriage airfoil was a hallmark of later Fokker designs.

The first Fokker V5 was given the military designation Fokker F.I 101/17; it was a refined V4 and the prototype of the Fokker Dr.I production triplane. F.I 101/17 was sent to Adlershof for structural testing on 7 August 1917, which it passed with minor issues that were easily corrected. Fokker F.I 102/17 and 103/17, which had been under construction since 11 July, were intended for front-line evaluation and arrived at *JG*I, the famous Flying Circus, on 28 August. F.I 102/17 was allocated to *Jasta* 11 for use by Manfred von Richthofen and the *Staffelführer*, *Lt.* Kurt Wolff, while F.I 103/17 was allocated to *Jasta* 10 for its *Staffelführer*, *Lt.* Werner Voss.

Richthofen scored his first victory in F.I 102/17 on 1 September and his second on 3 September.

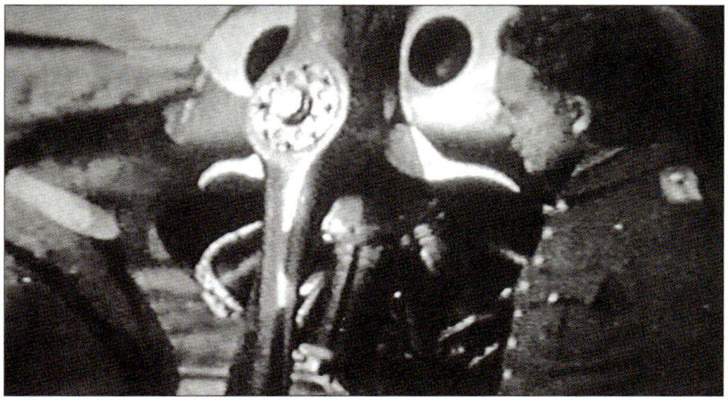

Voss also scored victories in F.I 103/17 but details are not available. The first triplane loss was on 13 September when *Oblt.* Kurt Wolff, who had been promoted on the 10th, was shot down and killed while flying F.I 102/17 during combat with a number of Sopwith Camels.

Voss returned to the front after a short leave and on 23 September shot down his 48th aircraft while flying F.I 103/17. Later that day he was shot down and killed in combat with seven pilots of 56 Squadron, RFC, most of whom were aces. This became the most celebrated dogfight of the war, and highlighted the Fokker Triplane's exceptional maneuverability and excellent climb rate. During the combat Voss easily out-maneuvered the pilots flying their S.E.5a fighters and was able to out-climb them at will. Voss repeatedly climbed above the British fighters during the combat and could easily have disengaged, but he dove back into the fight and continued against severe odds until he was finally killed.

Within 10 days both Fokker Triplanes had been lost and their ace pilots killed in action against overwhelming odds. The pilots may have been too

Left: Werner Voss in front of his Fokker F.I 103/17 showing his personal marking, a face. Historians have tried for years to determine if the cowling was standard Fokker olive or the *Jasta* 10 unit color of yellow.

Frontbeststand Inventory of Dr-Type Aircraft (Triplane Fighters) at the Front

	1914			1915						1916						1917						1918			
Manufacturer and Type	31 Aug	31 Oct	31 Dec	28 Feb	30 Apr	30 Jun	31 Aug	31 Oct	31 Dec	28 Feb	30 Apr	30 Jun	31 Aug	31 Oct	31 Dec	28 Feb	30 Apr	30 Jun	31 Aug	31 Oct	31 Dec	28 Feb	30 Apr	30 Jun	31 Aug
Fokker Dr.I																			2	17	35	143	171	118	65
Pfalz Dr.I																							9	7	1
Total:																			2	17	35	143	180	125	66

impetuous to attack against such great numerical odds, perhaps due in part to their confidence in the Triplane's capabilities.

The first production Fokker Triplane to be sent to the front was Dr.I 115/17 on 4 October 1917. This aircraft was sent to *Lt.* Heinrich Gontermann, *Staffelführer* of *Jasta* 15. Due to illness Gontermann did not fly the airplane until 26 October. Gontermann considered the Triplane to be "fabulous" but "in any case, with mine I will take my time and feel my way with caution." He had learned the lesson from Wolff's and Voss's deaths.

This first production aircraft was soon followed by 17 more that arrived in late October, then a subsequent batch of five before the end of the month for delivery to *Jasta* 11, the unit that Richthofen flew with while commanding *Jagdgeschwader* I, his famous "Flying Circus".

The introduction of the production Triplane into service was traumatic. *Vzfw.* Lautenschlaeger was shot down by mistake while flying Dr.I 113/17, and Richthofen wrote off 114/17 after an emergency landing due to engine failure. Much worse, on 30 October *Lt.* Gontermann was tragically killed when he crashed due to wing failure, and the next day *Lt.* Pastor was killed when the top wing of 121/17 failed in the air. The Triplane was grounded and an investigation was started.

In retrospect a number of technical faults, perhaps combined with mediocre quality control during construction, caused the failures of the upper wing and ailerons. The wings of the existing airplanes had to be replaced with new wings. Assembly procedures of the new wings were made more rigorous, the ailerons were modified for additional strength, and strengthened wing ribs were used. The Triplane was generally a robust aircraft, but despite the new, strengthened wings, at least three more episodes of wing failure occurred, fortunately without any further fatalities.

Meanwhile, another triplane, the Pfalz Dr.I, was type-tested at Adlershof in October 1917. The Pfalz was not quite as maneuverable as the Fokker, but powered by the 160 hp Siemens-Halske Sh.III rotary, its advantage was an exceptional climb rate. It was planned to build 100 of this new fighter and introduce them en masse to the front in January 1918. However, this was impossible due to the factory's inability to produce enough engines, which were still not fully developed for production, and only ten Pfalz Triplanes could be built. Nine were recorded at the front at the end of April, and one was photographed at *Jasta* 73, but nothing more is known about the combat career of the Pfalz Dr.I.

Rotary engine shortages, coupled with the need to replace all the top wings in the initial production aircraft, delayed getting the Fokker Triplane into service at the front. After the botched start, Triplane production never caught up with front-line demands and the twelve *Staffels* of *JG*I, *JG*II, and *JG*III were never fully equipped with the Triplane as intended. It was planned that these three elite units would spearhead the German air effort during the massive spring offensive, Operation Michael, scheduled for 21 March 1918. Completely equipping them with

Fokker Dr.I Specifications

Engine:	110 hp Oberursel Ur.II	
Wing:	Span Upper	7.190 m
	Span Middle	6.225 m
	Span Lower	5.725 m
	Wing Area	18.66 m^2
General:	Length	5.77 m
	Height	2.95 m
	Empty Weight	406 kg
	Loaded Weight	586 kg
Maximum Speed:		185 km/h
Climb:	1000m	2.9 min
	2000m	5.5 min
	3000m	9.3 min
	4000m	13.9 min
	5000m	21.9 min
Ceiling:		6000m
Duration:		1.5 hrs
Armament:		2 mgs

Pfalz Dr.I Climb (Full Specifications on Page 20)

Climb:	1000m	1.5 min
	2000m	3.0 min
	3000m	5.5 min
	4000m	8.5 min
	5000m	11.5 min
	6000m	14.5 min

Above: Reproduction Fokker Triplane in authentic camouflage and markings for Lothar von Richthofen of *Jasta* 11. Lothar, a 40-victory ace, was the Red Baron's younger brother. The red struts, wheel covers, and cowling are the *Jasta* 11 markings; the yellow tail and top wing are Lothar's personal markings.

Left: Manfred (at right) and Lothar (at left) von Richthofen in March 1918. Both are wearing the *Pour le Mérite*, also known as the Blue Max, around their necks.

the best fighter available, the Fokker Triplane, was part of the plan.

From a strategic standpoint, the German spring offensive, better known as the *Kaiserschlacht* (the Kaiser's battle), was Germany's last chance to win the war on the Western Front before the arrival of masses of American troops. This emphasized how critical its success was, and how important it was that the German fighter units have the best airplanes possible. Unfortunately for the Germans, the fighter they really needed, the Fokker D.VII biplane with high compression Mercedes or BMW engines, arrived at the front too late. By the time the D.VII arrived in quantity, the German offensive had failed and the momentum had shifted irrevocably to the Allies.

After the rocky start to its service at the front, Fokker Triplane numbers started to increase and peaked during the period February through May 1918, coinciding with the *Kaiserschlacht* and its aftermath. By then the Triplane's shortcomings were

Above: The Fokker Dr.i is one of the most famous airplanes of WWI; this is a beautiful flying reproduction of a *Jasta* 11 Triplane formerly flown by *Lt.* Eberhard Mohnicke, who was wounded on 1 March 1918. The swastika was a Nordic good-luck symbol before its post-war adoption by the National Socialist Party.

Below: Manfred von Richthofen, third from right, views the new Fokker F.I 102/17 that has just arrived for him. The F.I triplanes were pre-production aircraft and differed slightly from the production Dr.I. The production Dr.I had skids under the lower wingtips to protect them during ground operations. The Dr.I also had straight leading edges on its horizontal stablizers; these leading edges were curved slightly on the F.I aircraft as can be seen here.

Above: The Fokker Dr.I fighters of Jasta Boelcke are lined up for their next mission. The aircraft in the foreground is Fokker Dr.I 213/17 flown by *Lt.d.R.* Friedrich Kempf; another view is shown on page 78.

Below: The Fokker Dr.I was in production in January 1918 to equip the *Jastas* with a new fighter for the coming spring offensive, and two were at the competition for comparison with the new fighters being evaluated. An original Triplane is shown in Sanke Card 1055 on a snowy airfield.

Fokker Dr.I 425/17 flown by Manfred von Richthofen, *JGI*, as delivered. The iron cross insignia were over-painted by the straight-sided cross insignia for Operation Michael.

Fokker Dr.I 152/17 flown by Manfred von Richthofen, *JGI*.

Fokker Dr.I flown by *Lt*. Hans Joachim Wolff of *Jasta* 11, March 1918. Wolff scored 10 victories before he was killed in action 16 May 1918.

Fokker Dr.I 155/17 flown by *Lt*. Eberhard Mohnicke of *Jasta* 11, March 1918. Mohnicke scored 9 victories and survived the war.

Above: *Hptm*. Wilhelm Reinhard by his Fokker Triplane while serving with *Jasta* 11. He became commander of the Flying Circus after Richthofen was killed.

Right: Reinhard suffered partial upper wing failure while flying his Triplane, but appears undeterred. However, it was a portent of things to come. While testing the Zeppelin-Lindau D.I all-metal fighter on 3 July 1918, Reinhard was killed when the upper wing broke off. With 20 victories he had been nominated for the *Pour le Mérite*, but his accidental death prevented the award.

Fokker Dr.I 454/17 flown by *Lt.* Lothar von Richthofen of *Jasta* 11, March 1918. Lothar scored 40 victories, received the *Pour le Mérite*, and survived the war.

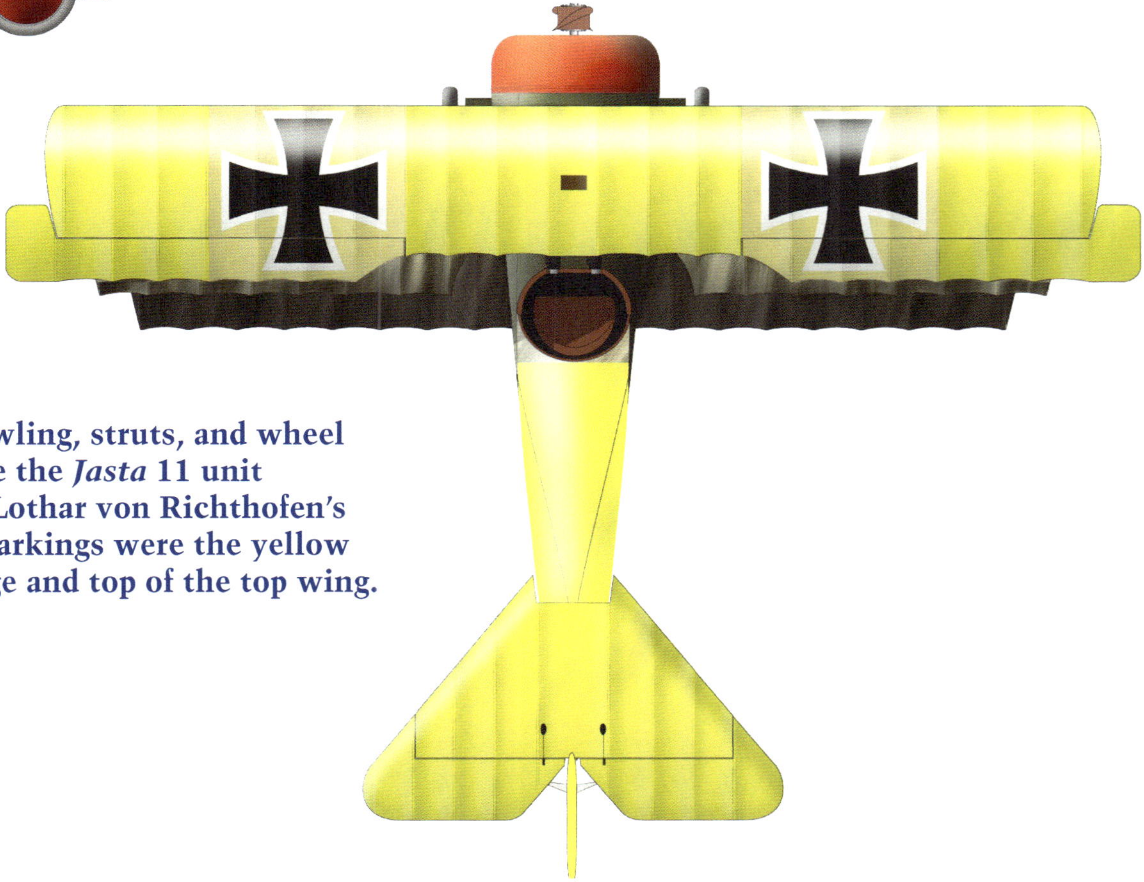

The red cowling, struts, and wheel covers were the *Jasta* 11 unit markings. Lothar von Richthofen's personal markings were the yellow rear fuselage and top of the top wing.

Fokker Dr.I 163/17 flown by *Lt.* Hans Karl von Linsingen of *Jasta* 11, January 1918.

Above: Fokker Dr.I photographed with its essential ground crew.

Right: Lineup of the Fokker Triplanes flown by *Jasta* 11.

Below: Manfred von Richthofen at left plays with his dog Moritz while talking with Erich Löwenhardt at right. Löwenhardt scored 53 victories to become Germany's 3rd ranking ace before being killed in action.

Above: Reproduction Fokker Triplane in authentic camouflage and markings for Lothar von Richthofen of *Jasta* 11.

clear. At low altitude the Triplane could still use its superior maneuverability and climb, but it was not fast enough at high altitude and its useful ceiling was inadequate. The increasing performance of Allied airplanes had now put a premium on ceiling and speed at high altitude. However, pilots generally preferred the Triplane to the outdated Albatros and Pfalz fighters that were the alternatives. By 1918 the Albatros and Pfalz were inferior to most of the Allied fighters in all respects, whereas the Triplane was at least superior in climb and maneuverability, and pilots preferred having some advantages to no advantages.

The Fokker Triplane was always out-numbered at

Fokker Dr.I 503/17 had the streaked factory camouflage finish with the *Jasta* 19 unit markings of white cowling and black and yellow stripes on the tailplane. *Lt.* Hans Körner's personal marking is the white lightning bolt on the sides and turtledeck. Körner scored 7 victories and survived the war.

Above & Below: Pfalz Dr.I 222/17 is shown here at *Jasta* 73 at Mars sous Bourq airfield without armament. The delivery pilot was Lehmann, shown in the cockpit below. Part of the lower cowling is cut away for additional cooling of the slow-turning 160 hp Siemens & Halske Sh.III.

Above & Below: Pfalz Dr.I 222/17 was one of ten pre-production triplanes ordered in October 1917. It is shown here at *Jasta* 73 at Mars sous Bourq airfield without armament. It was flown primarily by *Ltn*. Fritz Anders, the *Jasta* CO. These photos were taken after mid-March 1918 because the Albatros in the background has *Balkankreuz* markings, which were not introduced until that time.

Pfalz Dr.I 222/17, *Jasta* 73

Above: Pfalz Dr.I 221/17 with lower cowling cut away.

Above: Armed Pfalz Dr.I at Adlershof.

Fokker Dr.I 502/17 of *Jasta* 11, March 1918.

Fokker Dr.I 436/17 flown by *Lt.d.R.* Fritz Hoffman of *Jasta* 12, March 1918. Hoffman scored 1 victory before dying of wounds suffered in air combat.

Fokker Dr.I flown by *Lt.* Hermann Becker, *Staffelführer* of *Jasta* 12. Becker scored 23 victories and was nominated for the *Pour le Mérite* but did not receive it due to the Kaiser's abdication.

Fokker Dr.I 217/17 flown by *Oblt.* Paul Blumenbach of *Jasta* 12, March 1918. Blumenbach scored 3 victories and commanded *Jasta* 12 from March to 15 May.

Above: The most famous airplane of WWI, Dr.I 425/17 was the all-red Fokker flown by Manfred von Richthofen when he was killed in action on 21 April 1918. Examination of fabric from this aircraft reveals no camouflage, indicating it was painted for the Red Baron at the Fokker factory. A reproduction of this aircraft is shown on page 5.

the front by the old standbys, the Albatros D.V and Pfalz D.III. Only 320 Triplanes were ordered because newer, more effective fighters were in the offing. Once the Fokker D.VII started to arrive in quantity in May, the Triplane's days were numbered. The D.VII was not as maneuverable as the Triplane but was faster, had a higher useful ceiling, was easier to fly, and was more reliable. The D.VII started replacing the Triplane, the Albatros, and the Pfalz once it arrived at the front. On 24 June the *Kofl* of the XVIII *Armee* ordered the Triplanes of *JG*II removed from service. Similarly, by mid-June most of the Triplanes in use by *JG*III has been sent back to *Armee Flugpark* VII. However, a few Triplanes lasted in *Jasta* 36 of *JG*III into late July. When D.VII fighters with high-compression engines started arriving in June, their climb, ceiling, speed, and maneuverability at altitude established superiority over all other fighters at the front.

Once the *JG*s started to release their Fokker Triplanes, they became available for use by other *Jagdstaffeln*. Some Triplanes were put to good use for a time, but most were replaced by the D.VII as soon as production of the Fokker biplane permitted. A few Triplanes lingered on at the front, often used by aces for low-altitude combat, until the war's end.

Above: Fokker F.I 103/17 of Werner Voss, the first Fokker Triplane pilot to fall in combat.

Below: Fokker Dr.I 404/17 of *Hptm*. Adolf von Tutschek, *JG*II, March 1918. After scoring 27 victories Tutscheck was killed in action while flying this aircraft on 15 March 1918.

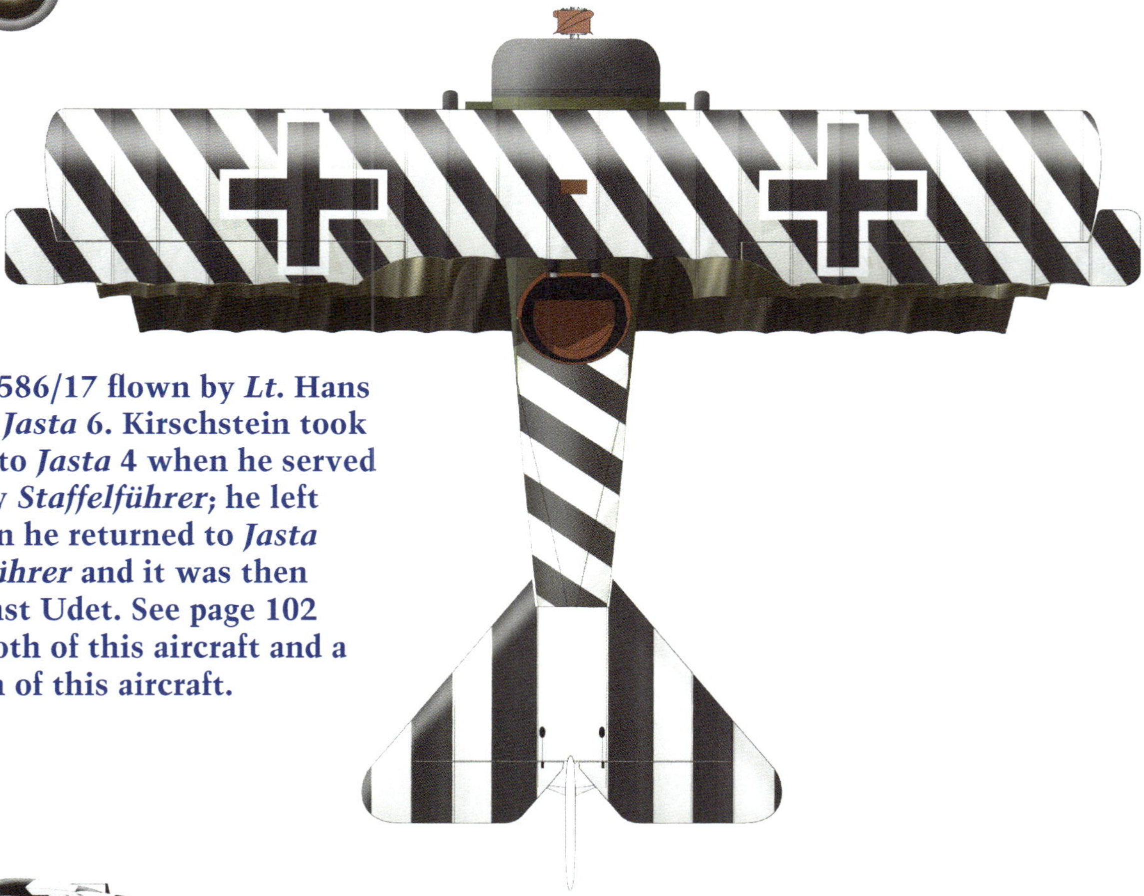

Fokker Dr.I 586/17 flown by *Lt.* Hans Kirschstein, *Jasta* 6. Kirschstein took this aircraft to *Jasta* 4 when he served as temporary *Staffelführer*; he left it there when he returned to *Jasta* 6 as *Staffelführer* and it was then flown by Ernst Udet. See page 102 for photos both of this aircraft and a reproduction of this aircraft.

Fokker Dr.I 586/17 as flown by *Lt.* Ernst Udet, who added his fiancee's initials to the side of the aircraft. Udet scored 62 victories and was awarded the *Pour le Mérite*.

Above & Left: The USAF Museum's reproduction Fokker Dr.I 503/17 in the markings of *Lt.* Arthur Rahn, a six-victory ace of *Jasta* 19, as of May 1918. The black and yellow horizontal tail markings and white cowl were the *Jasta* 19 unit markings, and the band of diamonds around the fuselage was Rahn's personal marking.

Above: Fokker Dr.I 146/17 photographed at *Jasta* 11 in April 1918. The rear fuselage and tailplane have been painted along with a fuselage band for pilot identification.

Above: Fokker Dr.I 127/17 soon after arriving at Jasta 11 before unit markings were added. Manfred von Richthofen scored his 71st, 74th, and 76th victories flying this Triplane.

Below: *Jasta* 19 lineup with *Lt.* Rahn's aircraft at right.

Fokker Dr.I flown by *Vzfw.* Ulrich Neckel of *Jasta* 12, March 1918. Neckel scored 30 victories, was promoted to *Leutnant*, and received the *Pour le Mérite*. Neckel survived the war.

Fokker Dr.I of *Jasta* 12

Fokker Dr.I 193/17 flown by *Lt.* Wilhelm Schwartz, *Staffelführer* of *Jasta* 13. May 1918. Schwartz scored 8 victories.

Fokker Dr.I 183/17 flown by *Oblt.* Hasso von Wedel, *Jasta* 14. Wedel scored 5 victories and later commanded *Jasta* 75, then *Jasta* 24 and *Jagdgruppe* 12.

Above: Fokker Dr.I 213/17 flown by *Lt.d.R.* Friedrich Kempf of *Jasta* Boelcke (*Jasta* 2) in March 1918. "Kennscht mi noch?" means "Remember me still?" Kempf scored four victories and survived the war. The tailplane has been painted in *Jasta* Boelcke's unit markings of black and white like the adjacent profile.

Fokker Dr.I flown by *Lt.d.R.* Richard Krüger of *Jasta* 4. Krüger scored one victory and died of wounds from air combat.

Fokker Dr.I 471/17 flown by *Lt.* Hans Pippart, *Staffelführer* of *Jasta* 19, May 1918. Pippart scored 22 victories and was KIA on 11 August 1918.

Facing Page: *Lt.* Hans Pippart, *Staffelführer* of *Jasta* 19, with his Fokker Dr.I 471/17.

Fokker Dr.I 198/17 flown by *Lt.* Hans Werner of *Jasta* 14, January 1918. Werner scored 7 victories.

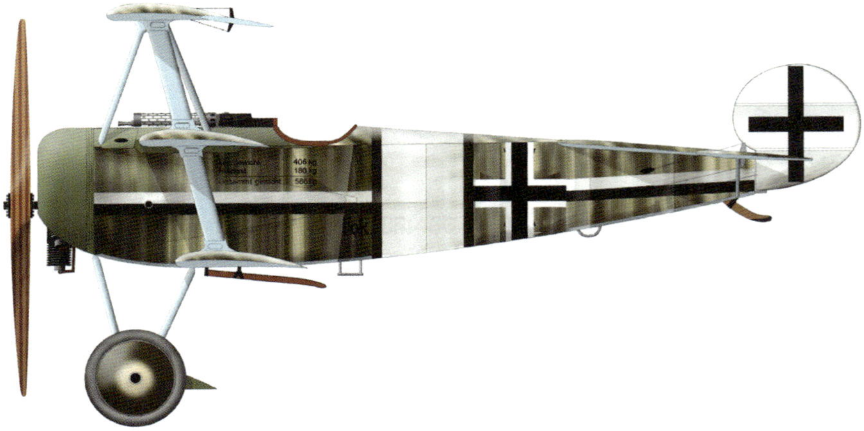

Fokker Dr.I 466/17 of *Jasta* 14.

Fokker Dr.I 588/17 flown by *Gefr.* Reinhold Preiss, *Jasta* 14, June 1918.

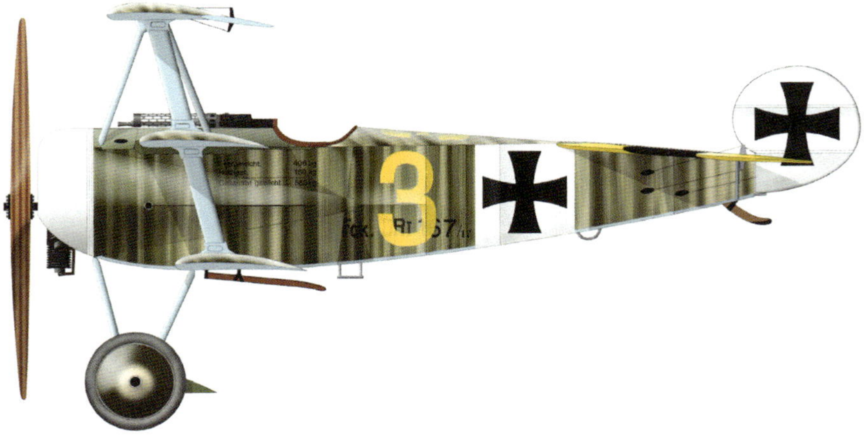

Fokker Dr.I 167/17 of *Jasta* 19

Above: Von Richthofen examines Fokker F.I 102/17 that has just been delivered to *JGI* for his use.

Below: Hermann Göring, center, with pilots of *JGI*. Göring became commander of *JGI* after Reinhard was killed. An ace with22 victories, Göring later became notorious as *Reichsmarschall* after Hitler achieved power in Germany.

Above & Below: Fokker Triplanes of *JG*II.

Fokker Dr.I 214/17 flown by *Lt.d.R.* Wilhelm Papenmeyer, *Jasta* Boelcke, March 1918. Papenmeyer scored 4 victories and was KIA.

Fokker Dr.I 419/17 flown by *Lt.* Walter Göttsch, *Staffelführer* of *Jasta* 19. Göttsch scored 20 victories and was KIA on 10 April 1918.

In Retrospect

Although the Sopwith Triplane was well-liked by its pilots and established a good record in combat with German aircraft, its most significant contribution to the British war effort may have been the amount of time and resources it caused the German aviation industry to waste on triplane fighter designs.

Significantly, the Sopwith company replaced their Triplane with a biplane, the famous Camel. And despite a number of further British triplane designs for fighters, bombers, and reconnaissance airplanes, none saw production and nearly all British warplanes that saw combat in WWI were biplanes. British authorities distrusted the structural strength of monoplanes and biplanes always seemed to out-perform triplanes.

German experience was similar although not identical. Unlike the Sopwith, which was as fast or faster than its German competitors when it arrived at the front, the Fokker Triplane was slower, as would be expected from the greater drag associated with triplanes. Where the Fokker Triplane excelled was great maneuverability and climb rate, both conferred by its triplane configuration.

One thing successful warriors learn from combat is the lessons their enemy teaches. And Allied fighter pilots learned to avoid dogfighting the Fokker Triplane when they could avoid it. Instead, Allied pilots relied on the greater speed of their Spads and SE.5a fighters to choose combat when it was to their advantage and avoid it when the Fokker Triplanes had a tactical advantage. Slower airplanes did not have this ability to choose whether or not to engage the Fokkers, and then the Fokker Triplanes climb, maneuverability, and fire-powered were used to excellent effect.

Nevertheless, the triplane fighter was an anomaly. In a world of biplanes, the triplane had the great disadvantage of being slower.

Ironically, Anthony Fokker, who designed the Fokker Triplane, led the return to the monoplane fighter in mid 1918. Fokker worked with a prominent German plywood manufacturer to develop the cantilever wing design that he used in his triplane. When the slow speed of the triplane became apparent, Fokker designed the D.VI biplane that was clearly a derivative of his Triplane. Although not quite as maneuverable as the Triplane because of its longer wingspan, the D.VI was faster; just what one would expect based on the inherent qualities of the two configurations.

Produced in small numbers due to its low power rotary engine, the D.VI was over-shadowed by the D.VII of similar configuration. The D.VII was a better-performing, more potent fighter at high altitude due to its engine, and so was chosen for mass production.

While the D.VII established itself in combat as the best all-around fighter of the war, Fokker continued to experiment by designing and testing many fighter designs. His work led to a number of monoplanes, a configuration made practical by the strong, cantilever wing design that first appeared in the Fokker Triplane. In the Fokker E.V (later re-designated D.VIII) parasol monoplane, Fokker created the simplest practical fighter and endowed it with excellent performance despite its low power. The D.VIII was not as maneuverable as the D.VI biplane of similar engine from which it was derived, but again it was faster due to its reduced drag. The monoplane, not the triplane, was the future of airplane design. Fokker, designer of the only German triplane fighter to see wide-spread combat, also created the monoplane fighters that fore-shadowed the future. Although many biplanes were built in the 1920s and 1930s, by the second half of the 1930s the monoplane was clearly the way forward.

Today the biplane is archaic and the triplane seems somewhat humorous. Nevertheless – as may be seen from the photos of all the flying reproductions in this book – as the famous mount of the Red Baron, it has captured our imaginations and epitomizes the first air war.

Above: Fokker Dr.I 479/17 of *Lt.* August Raben, *Jasta* 18. Raben was the *Staffelführer.* 'Raben' is German for 'raven', and the *Staffel* adopted a raven as their unit marking. This aircraft was red overall in front with a white tail. It was being flown by another unit when captured by the French as shown here; by this time *Jasta* 18 was flying Fokker D.VIIs.

Bibliography

Books

Franks, Norman, Bailey, Frank, and Guest, Russel, *Above the Lines*, Grub Street, 1993.

Franks, Norman, Bailey, Frank, and Duiven, Rick, *The Jasta Pilots*, Grub Street, 1996.

Gray, Peter, and Thetford, Owen, *German Aircraft of the First World War*, second revised edition, New York: Doubleday & Company, Inc., 1970.

Green, William, and Swanborough, Gordon, *The Complete Book of Fighters*, Smithmark Publishers, 1994.

Grosz, Peter M., Haddow, George, and Schiemer, Peter, *Austro-Hungarian Army Aircraft of World War One*, Flying Machines Press, 1993.

Herris, Jack, *Development of German Warplanes in WWI*, Aeronaut Books, 2012.

Herris, Jack, and Pearson, Bob, *Aircraft of World War I: 1914–1918*, London: Amber Books Ltd., 2010.

Herris, Jack, *Pfalz Aircraft of WWI*, Aeronaut Books, 2012.

Imrie, Alex, *The Fokker Triplane*, Arms and Armour Press, 1992.

Leaman, Paul, *Fokker Dr.I Triplane*, Air War Classics, 2003.

Articles

Grosz, Peter M., "Frontbestand" *WW1 Aero* No.107, Dec. 1985, p.60–66.

Grosz, Peter M., "Frontbestand" *WW1 Aero* No.108, Feb. 1986, p.66–69.

Philips, Terry H., "The Fokker V.7 Experimental Triplanes", *Over the Front*, Vol. 22 No.3, 2007.

Selected Index

Left: The Fokker D.VI was clearly a biplane derivative of the Fokker Triplane. As might be expected, the D.VI was not as maneuverable as the Triplane, but was faster. About 60 were built and served with German home defense units. A small number were also supplied to Austria-Hungary. The D.VI was a good fighter that just needed a more powerful rotary to be superior, but the engines it needed were not available in production quantities due to development problems. The inline-engine Fokker D.VII of similar configuration was used instead.

Right: Ironically, by mid 1918 German designers were turning to monoplane fighters like this Fokker E.V, later re-designated the D.VIII. Using the same engine as the Triplane and D.VI, the D.VIII was faster but not as maneuverable. Like the D.VI it needed a more powerful rotary.

Brandenburg L 16

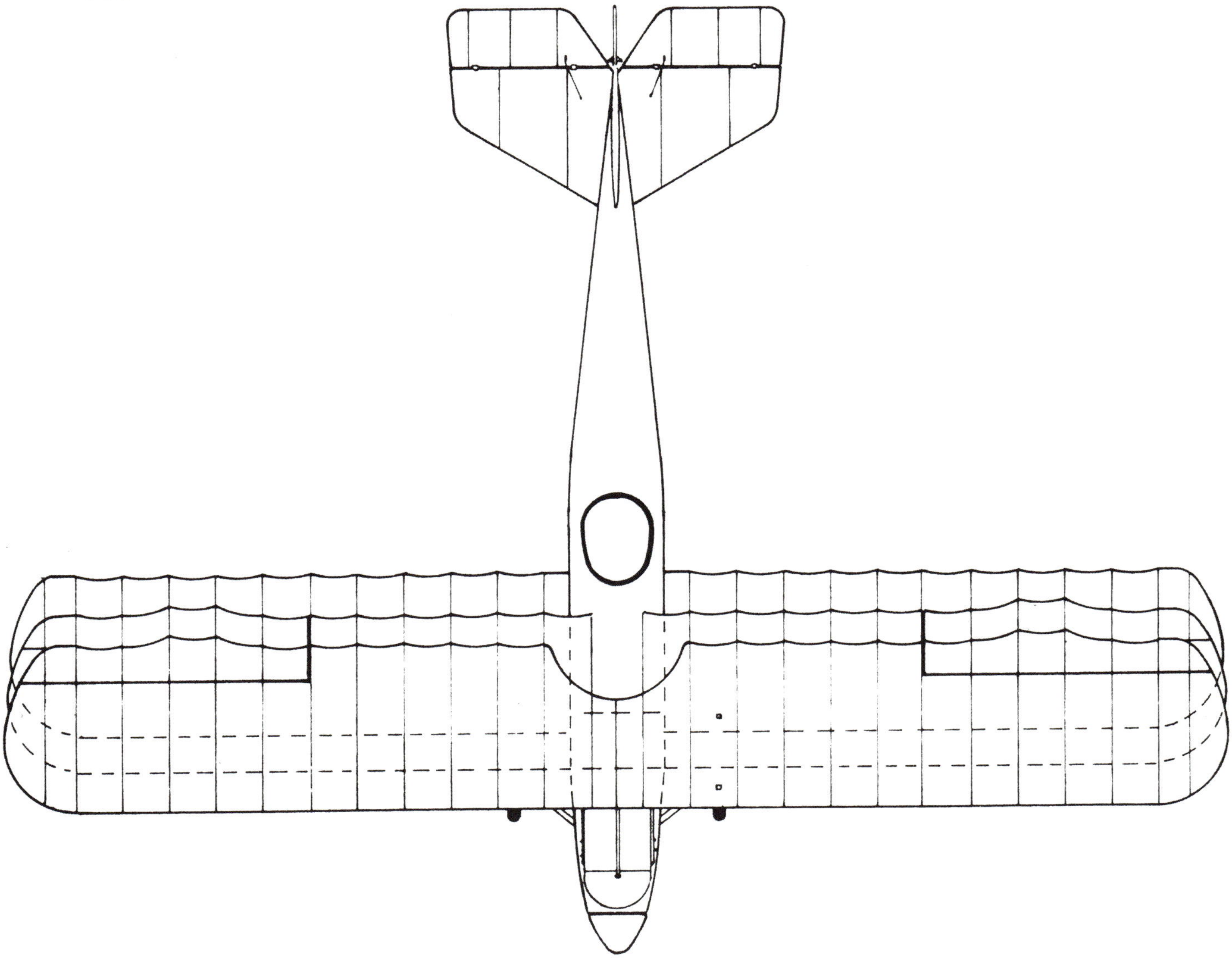

Brandenburg L 16

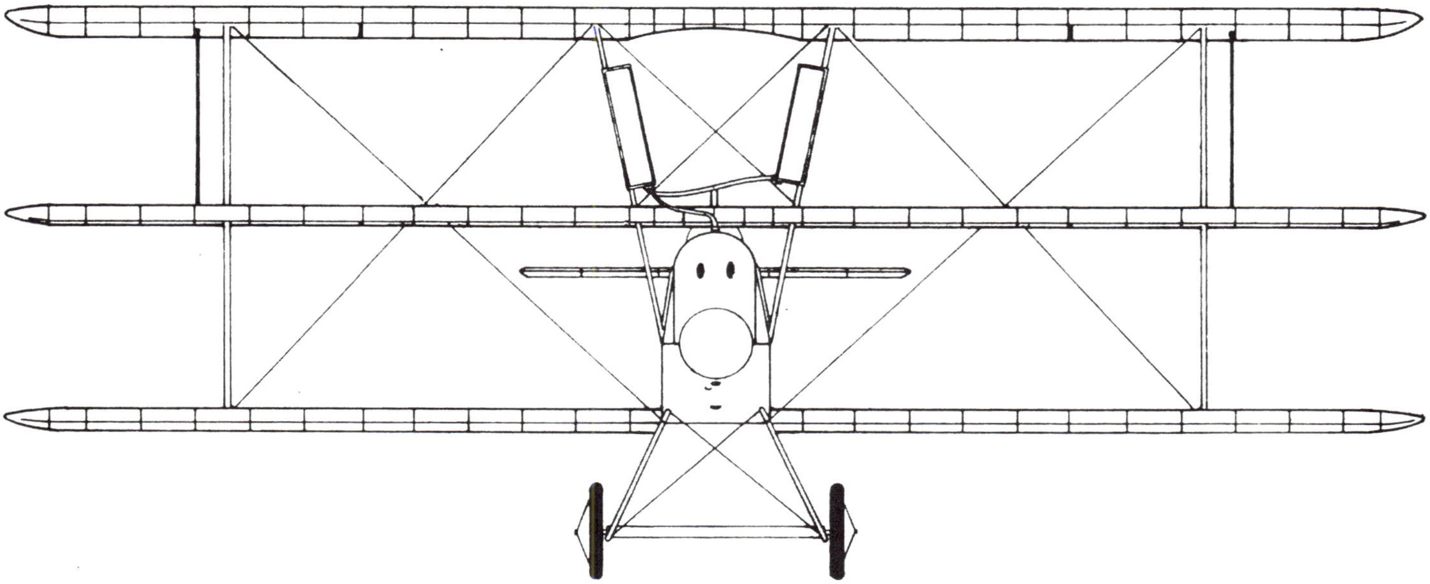

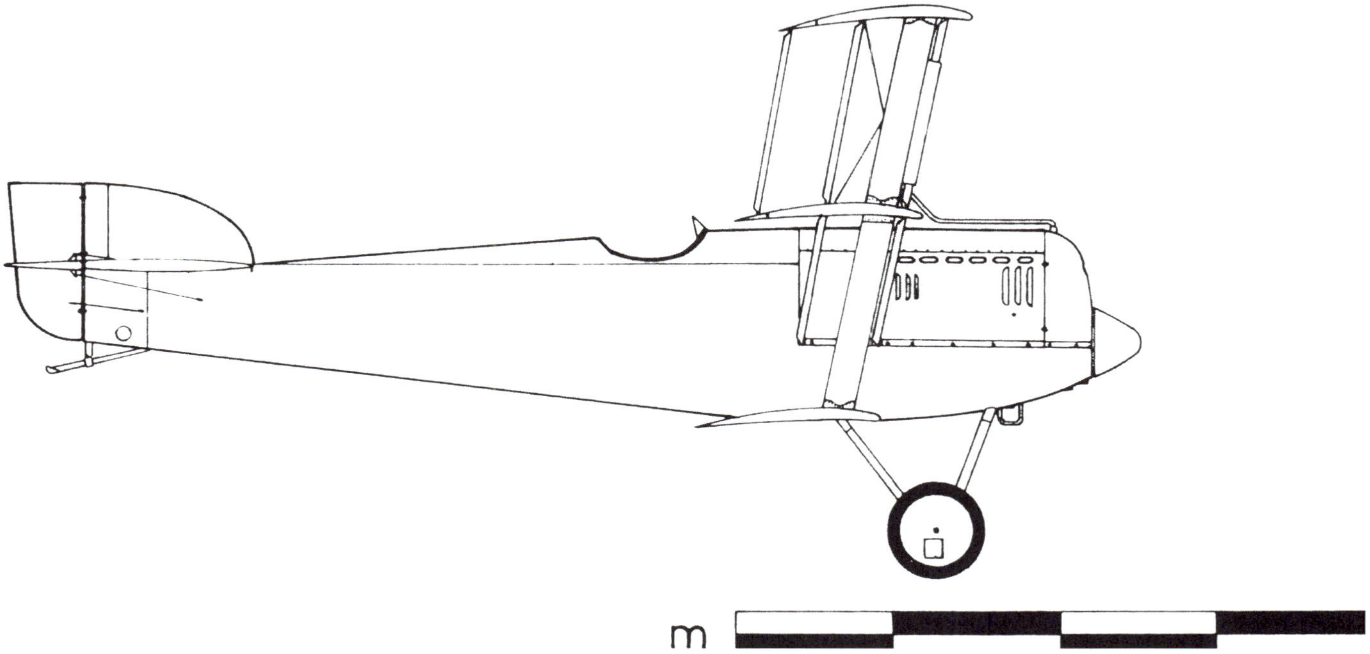

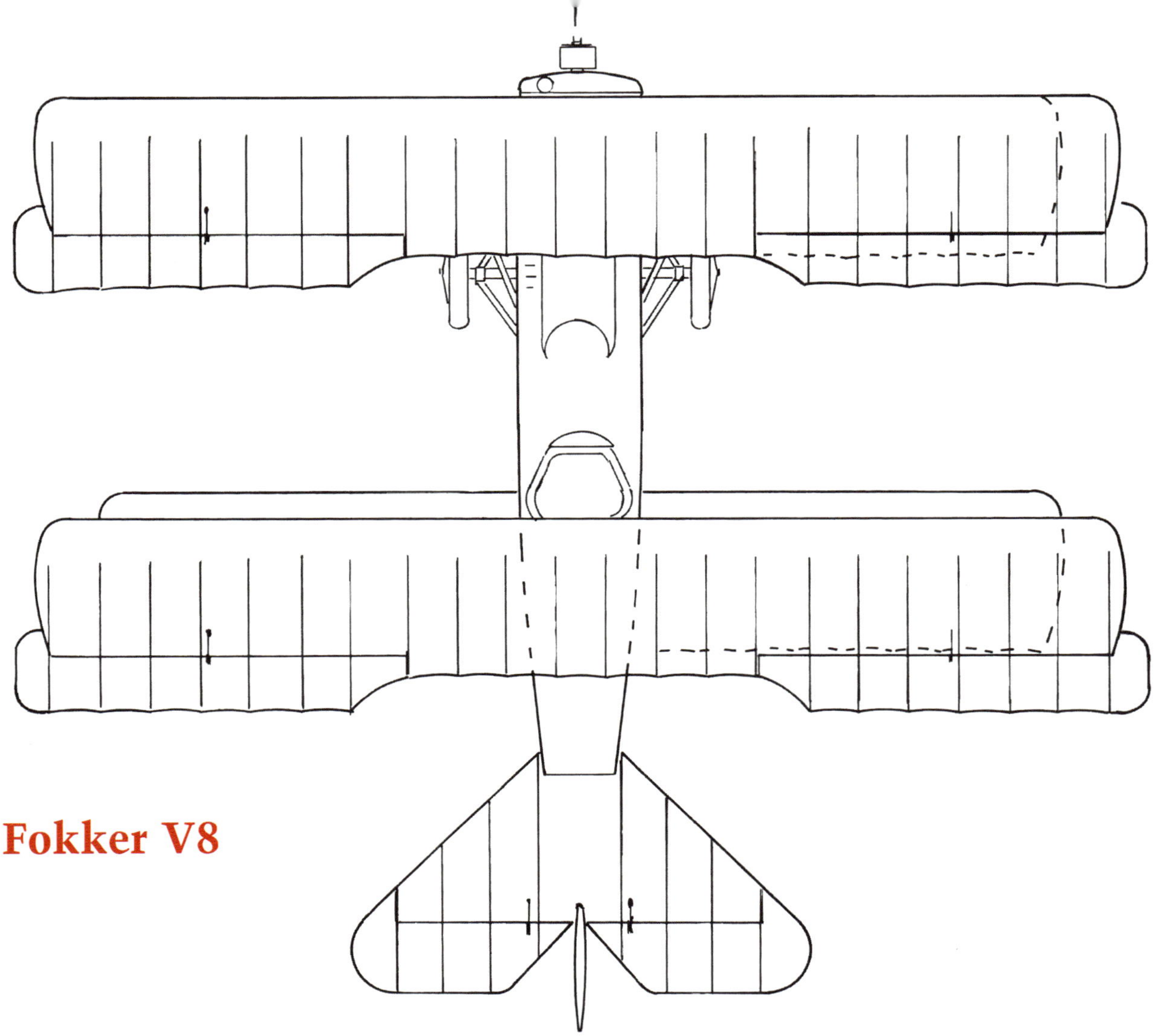

Fokker V8

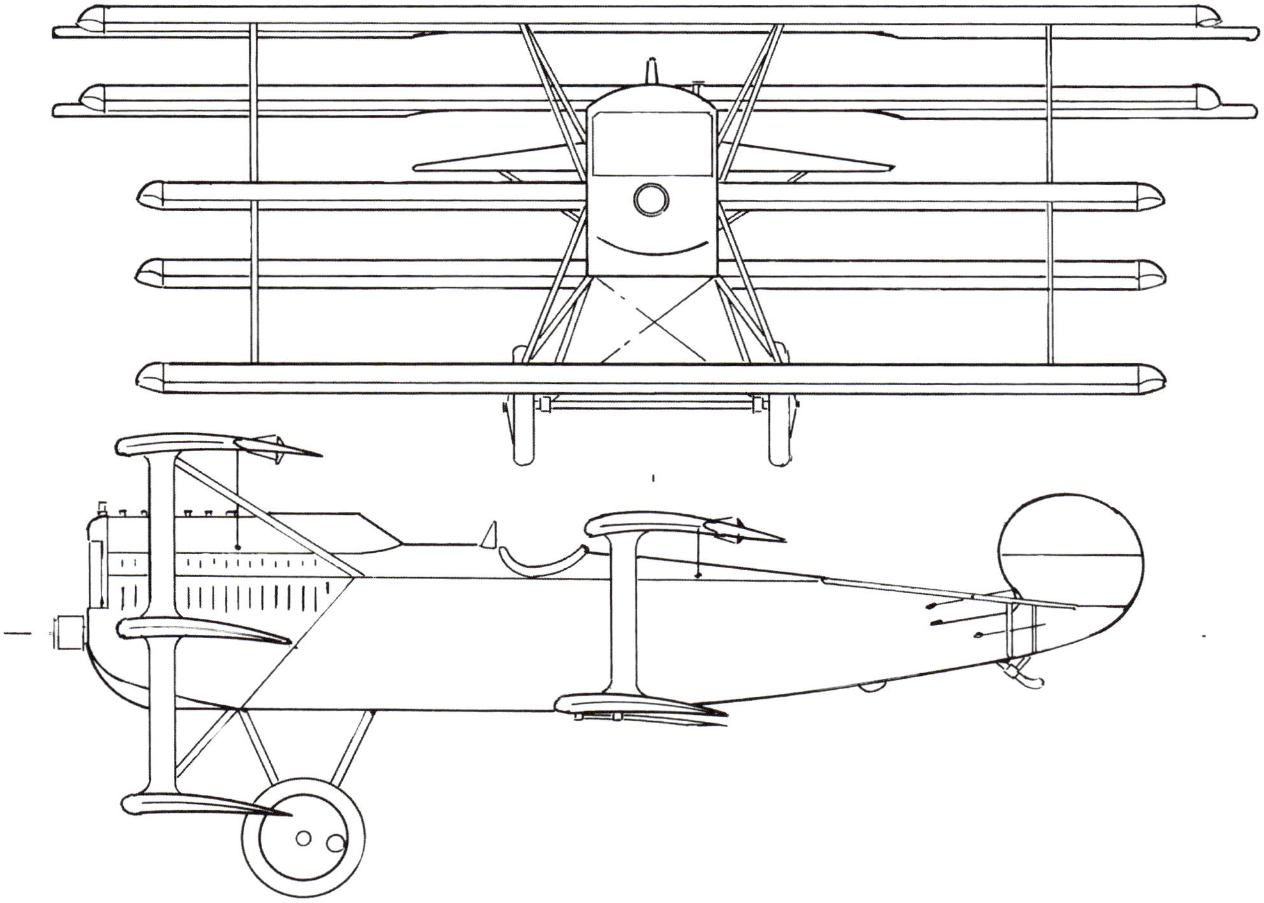

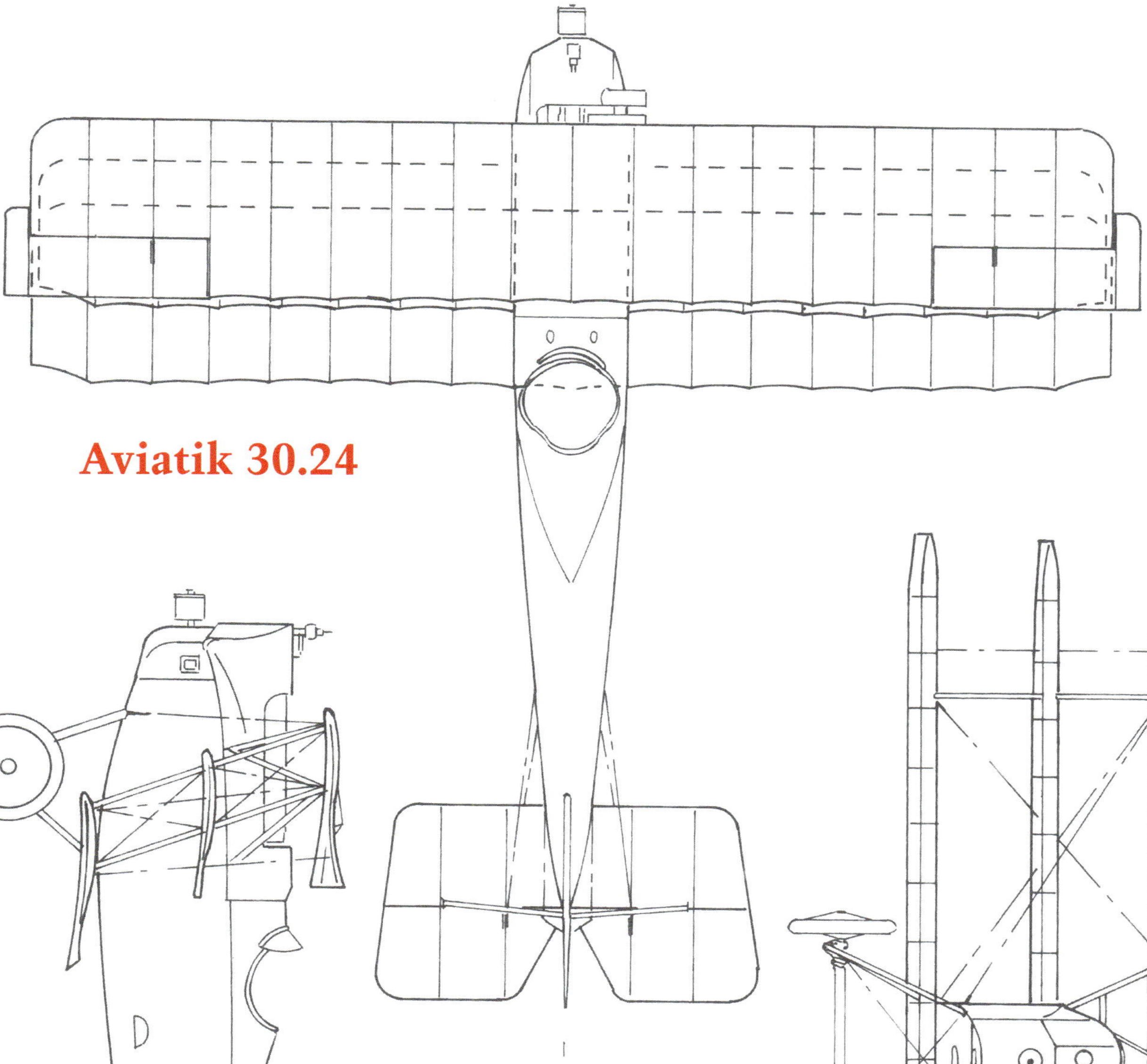

Aviatik 30.24

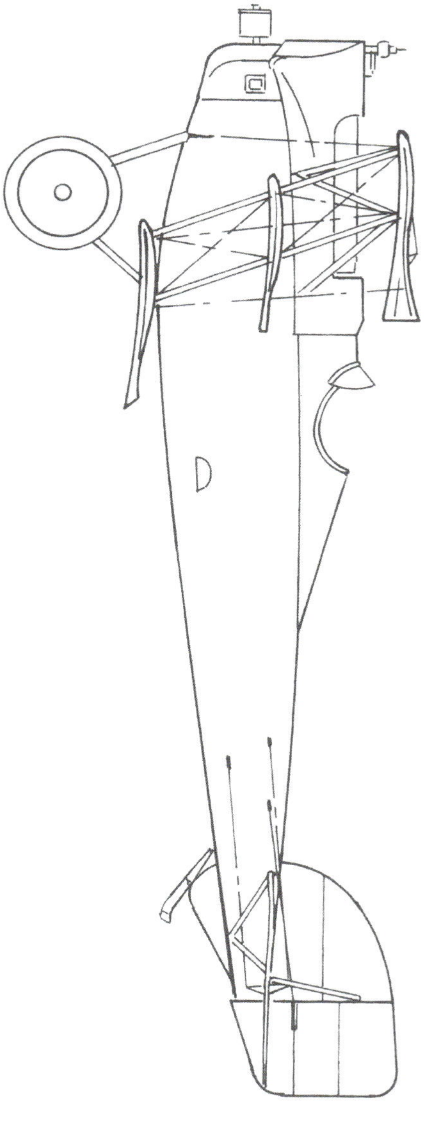

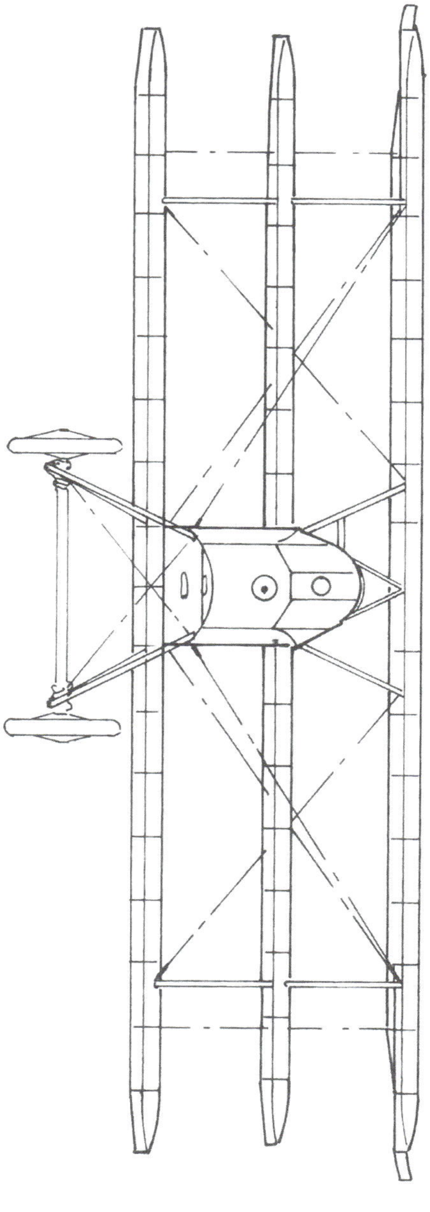

Friedrichshafen FF54

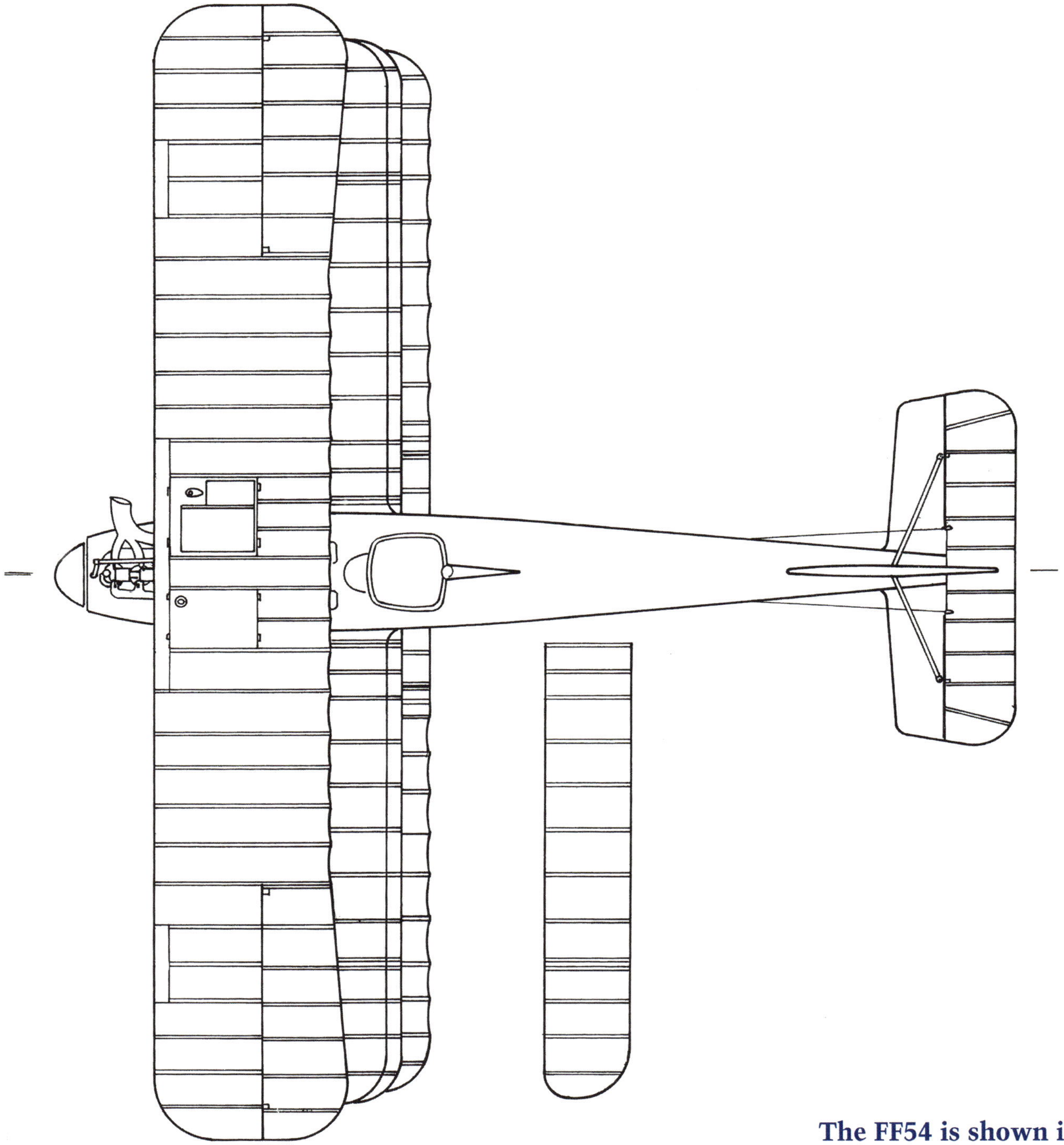

The FF54 is shown in its initial configuration as a quadraplane before being converted to a triplane.

Friedrichshafen FF54

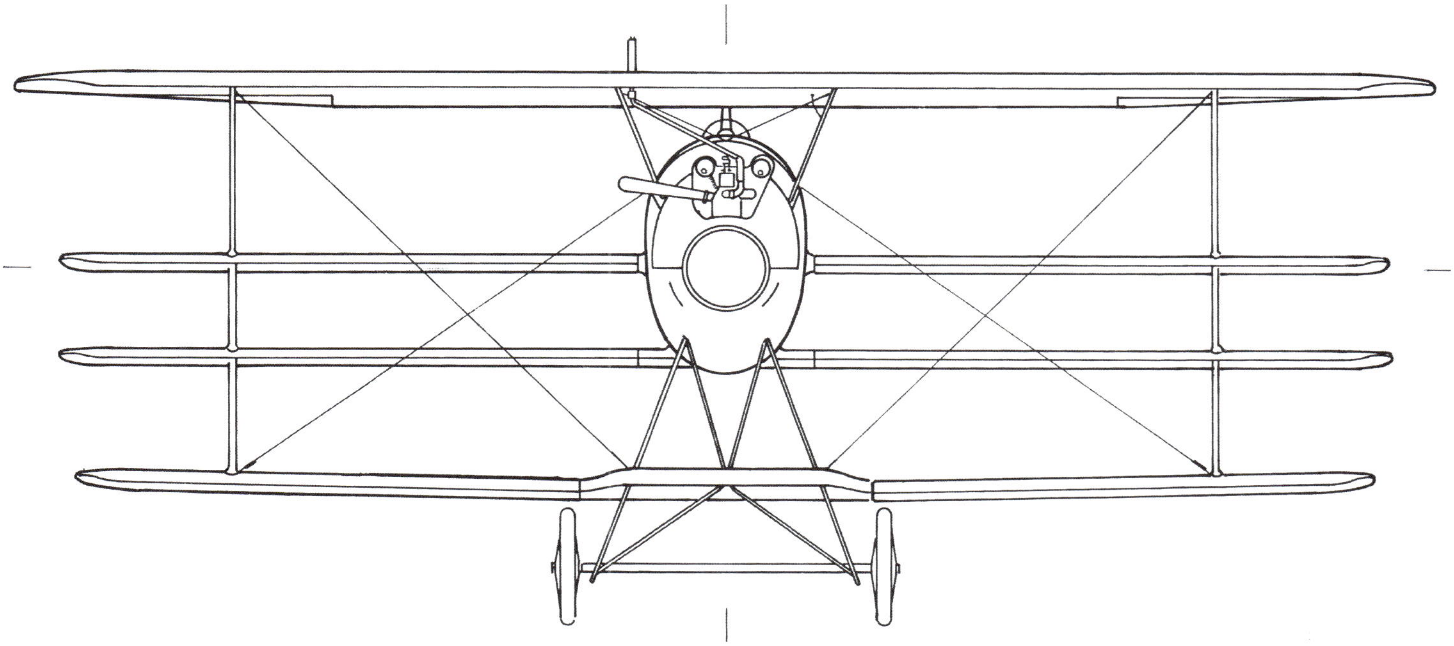

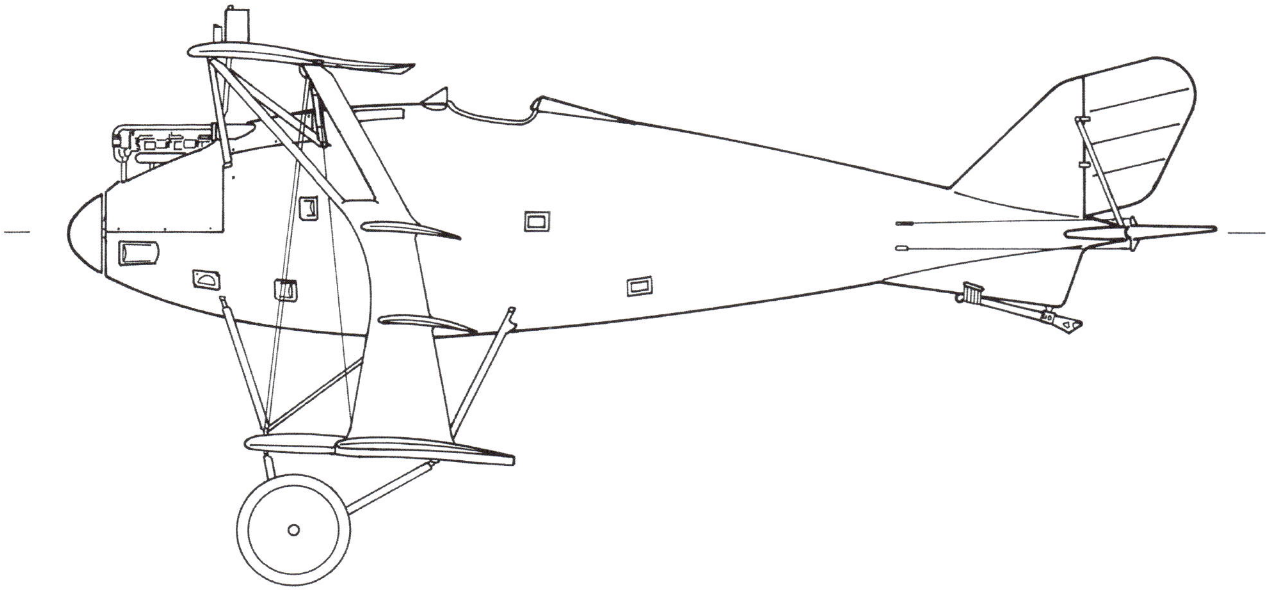

Schütte-Lanz Dr.I

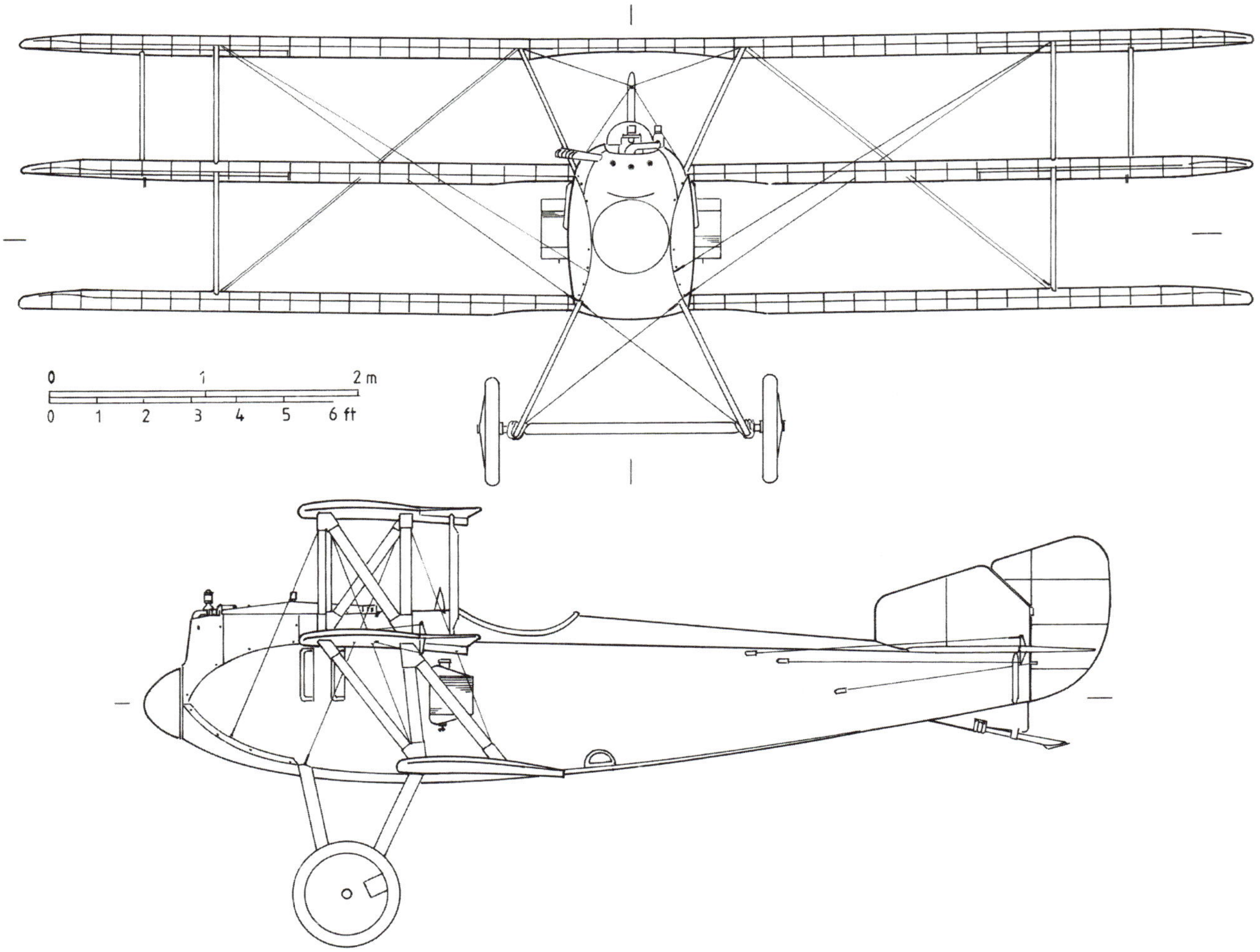

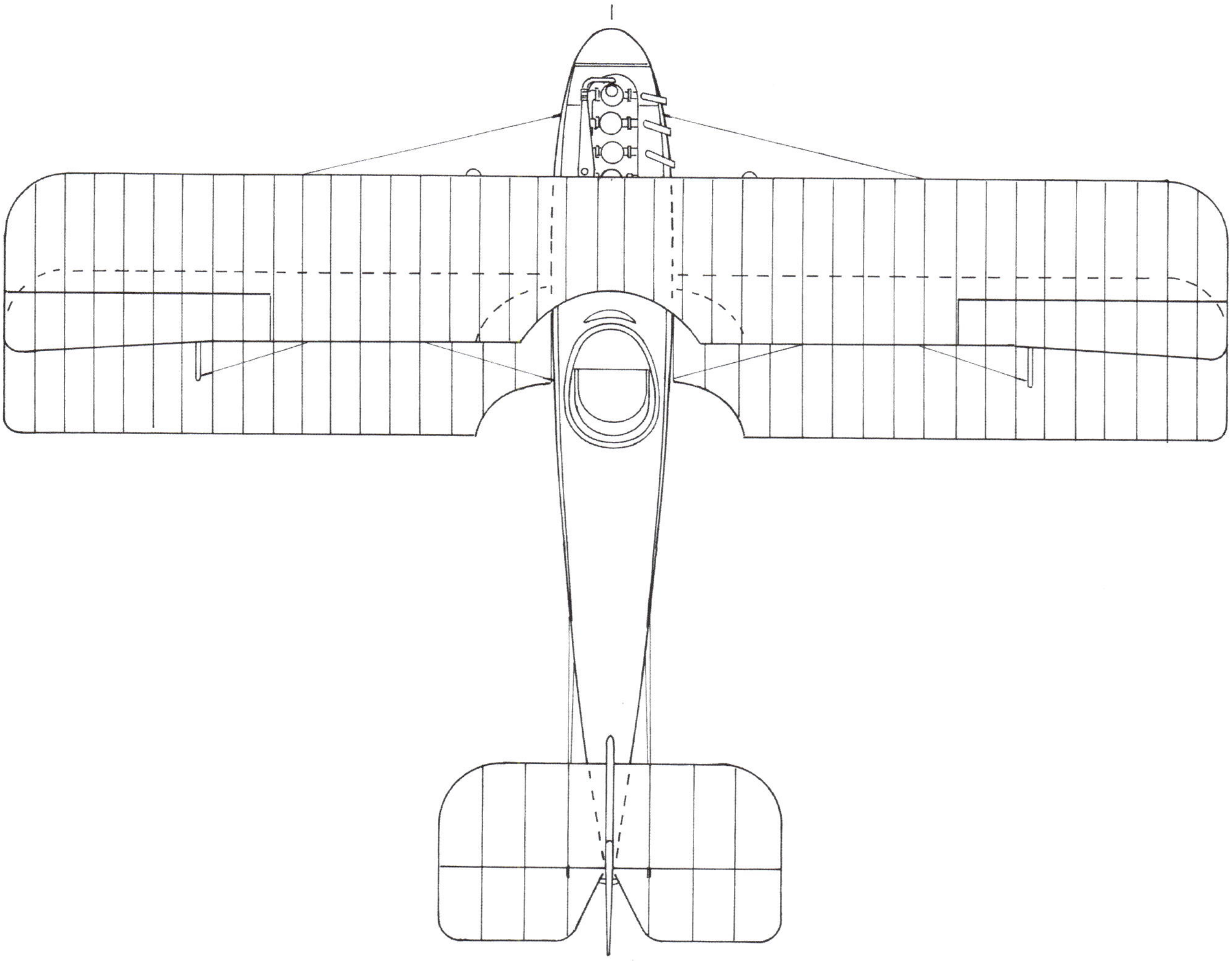

Afterword

When discussing Fokker Triplane aces, thoughts naturally go to Werner Voss and his epic combat with No.56 Squadron, and the most famous fighter pilot of them all, Manfred von Richthofen, the legendary Red Baron. However, great as these aces were, another German ace shot down more foes while flying the Fokker Triplane than either Richthofen or Voss.

Lt. Josef Jacobs achieved 48 confirmed victories, more than 30 of them while flying the Fokker Triplane. While Werner Voss also scored 48 victories, only a few were while flying the Triplane. The Red Baron scored 80 victoies, but only his last 21 were scored while flying the Triplane. This makes Josef Jacobs the leading exponent of the Fokker Triplane.

Jacobs learned to fly in 1912. When war broke out in 1914 he enlisted in the German Air Service. Following training as a military pilot, he was assigned to a two-seater unit, *FA*11. When the Fokker E.III fighter became available he flew those with Fokker Staffel West, which later formed the core of *Jasta* 12. However, Jacobs soon transferred to *Jasta* 22. After serving as a flight instructor at *Jastaschule* I, he joined *Jasta* 7 as its commander, where he served through the armistice.

Jacobs started flying Fokker Triplanes in early 1918 and continued to fly them into October 1918, long after the Triplane had been superceded by the Fokker D.VII, which he also flew. Jacobs flew at least two different Triplanes, which were painted black overall, and used them when poor weather limited the ceiling. The Triplane's exceptional climb and maneuverability at low level gave Jacobs the advantage in those circumstances. When the weather allowed flight at high altitudes, Jacobs flew his Fokker D.VII, which had a higher ceiling and better speed and climb at high altitude.

Above: *Lt.* Josef Jacobs was awarded the *Pour le Mérite* on July 18, 1918, and is shown here before receiving the award. Jacobs scored 48 confirmed victories, more than 30 of them while flying the Fokker Triplane, and survived the war.

Left: Fokker Dr.I 450/17 was one of *Lt.* Josef Jacobs' two personal Triplanes as commander of *Jasta* 7. It was overall black with a fire-breathing devil's head painted facing forward on both sides of the fuselage. Jacobs' more than 30 victories in the Triplane made him the leading Fokker Triplane ace. Flt. Cdr. Collishaw scored 33 victories flying the Sopwith Triplane.

Fokker Dr.I 450/17 flown by *Lt.* Josef Jacobs, *Staffelführer* of *Jasta* 7. Jacobs scored 48 victories and was awarded the *Pour le Mérite.*

Fokker Dr.I 504/17 flown by *Lt.* Rudolf Reinau of *Jasta* 19. Reinau scored 6 victories.

Above: Flying reproduction of the Fokker Dr.I 586/17 flown by *Lt.* Hans Kirschstein, *Jasta* 6. Kirschstein took this aircraft to *Jasta* 4 when he served as temporary *Staffelführer*; he left it there when he returned to *Jasta* 6 as *Staffelführer* and it was then flown by Ernst Udet. Kirschstein scored 27 victories before being killed in a flying accident as a passenger.

Below: *Lt.* Hans Kirschstein and his ground crew in front of his triplane, Fokker Dr.I 586/17.

20960150R00056

Made in the USA
Charleston, SC
30 July 2013